THE AUTOCAR COMPANY
(*) Special Items

CAR BUILDING RECORD

I0226650

Hercules	4"	REAR AXLE	4 TLG 0270	Brake Disc	3 D 0381	Battery	16 ZRM 0440		
ENGINE (JXD)	2 UG 080	5002-TW-X-1 Ratio 6.62	Brakes 16x3-1/2	Disc Brake Appl.	10 HA 0990	Location Install.	In Cab --		
Flywheel	With Engine 2 006	Gear Carrier	4 BA 0170	FRAME	12 UK 710 A	Tool Box	Location Installation	In Cab 16 --	
Pilot Brg.	3 A 0243	Brake Lining	4 -- 052	Frame (Frt. End) Assembly 12	-- No. 1	HOOD	16 UBK 0950		
Governor	2 UK 0650	Rr. Hub & Dr. Assem.	4 BP 0809	Frame (Center) Assembly 12	-- No. 2	HEADLIGHTS	16 UBK 0460		
Oil Filter	2 UBA 0808	Rr. Wheel Adapter	4 -- 0559	Frame (Rear End) Assembly 12	-- No. 3	Brakes	Air		
Clutch	3 RL 030 A	Rear Hub	4 BP 004	Undrilled Frame Side Rails	12 UK 0112 0113	Brake Sketch	2308		
Mfg. No. Long	12 CBC	Rr. Hub Cap Adptr.	4 --	Brake Cross Shaft Cross Member	12 UG 4480	Trailer Conn.	On Brake Sketch		
Carburetor Fuel Pump	6UG0510 UG 050	Rear Brake Drum	4 BP 021	SPRINGS, FRT.	13 UG 009	Hand Brake	H-15		
Air Cleaner United Oil Bath	ZBM 0330	Wheel Studs, R.H.	4 -- 0489	Frt. Spring Clip	13UGA042 13 UK 042	Air Gauge Install.	25UBM01010		
Air Cleaner Installation	6 UBK 0440	" " L.H.	4 -- 0491	" " Shim	13 UG 014	WHEELS, Frt. Size	20 x 8 L Rim		
Dist. & Gear	8 UG 0470	Drum Bolts	--	Thick End	Forward	" Part No.	9BJ0403		
Mfg. No. DR-EX-	37343	Axle & Flng. Assy.	4 -- 020	Frt. Spr. Keeper	13 UG 043	" Mfg. No.	33287		
Coil 6-V.	8 SA 0120A	Brake Type	Air	SPRS., REAR	13 UG 011	WHEELS, Rr. Size	20 x 8 L Rim		
Mfg. No.	DR-528C	RADIATOR	5 UK 590	Rear Spr. Clip	13 SDS 041	" Part No.	9BJ 0403		
Generator	16 BLF 0410	Fan (21")	5 UK 0120	" " Keeper	13 UG 092	" Mfg. No.	33287		
Mfg. No.	DR-1102404	6-Blade Fan Belt	5 UK 055	Rr. Spg. Clip Plt.	13 UG 0111	Spare Wheel & Tire	9BJ0403		
Voltage Regulator	16 BLG 0620	GASO. TANK	6 URK 4720	" " Shim	13 UG 0113-14	TIRES, Front	9.00-20		
Mfg. No.	DR-5540	Location	LH Rail	Thick End	Rear	" Make	--		
Starter	16 BLA 0560	Capacity	40 gallons	SPRGS., AUX.	13 UG 0156	TIRES, Rear	9.00-20		
Mfg. No.	DR-720T	Aux. Gaso. Tank	6 --	Brkt. Clearance	1-1/2 "	" Make			
MAIN TRANS.	8 UK 6530	Aux. Gaso. T'k Ins.	6 --	Aux. Spr. Brkts.	13 DK 0154	Tire Carrier	16 Omit		
Mfg. No. Clark	230F	Location	--	Run. Brd. & Spl. Grd. & Fdr. Assem., R.H.	14 -- 070	BUMPER	12UK402		
TRANSFER CASE	19 BK 0440	MUFFLER	7 TE 010	Run. Brd. & Spl. Grd. & Fdr. Assem., L.H.	14 -- 080	START. CRANK	1 ZN 010		
TIMKEN	T-2-B-7-4	Exhaust Pipe	7 UK 003	FENDER, R.H.	14 URK 052 A	Grille Radiator Guard	5URK4540 12 -- 0140		
Speedo Cable	16 LE 0800	Exhaust Pipe Ext.	7 -- 040	" L.H.	14 URK 053 A	Wheel Brg. Wrenches	Fr.— None Rr.— None		
" Adapter SK X Extension	SW 1.071	FRONT AXLE	9 BKW 6430	Spl. Guard, R.H.	14 URK 021	Wheel Wrenches	16SA0743-44		
" Gear	15T-205052	Ratio 6.62:1 Steering Arm	Brakes 16x3-1/2 9 With Axle	" L.H.	14 URK 022	Ignition Key No.			
Front Drive Shaft	30 LD 4110	Fr. Hub & Dr. Assem.	9 BP 0809	Runn. Bd.-R.H.	14 URK 343	Cab Door Key No.			
Series	1500	Frt. Hub	9 BP 004	Runn. Bd.-L.H.	14 URK 344	Tool Box Key No.			
Length	11-1/8	Frt. Hub Cap Adapter	9 --	" Iron, R.H.	14 -- 001	Chassis Wght., Frt.			
Front Axle Driveshaft	30 ZC 5330B	Frt. Brake Drum	9 BP 0232	" " L.H.	14 --	" " Rr.			
Series	1410	Brake Lining	9 --	" Brace (F)	14 -- 078	Tot. Chassis Wght.			
Length	33"								
Rear Driveshaft	30 DD 5340	Brake Type	Air	" " (R)	14 -- 078	Wheel Cuts	R.H. L.H.		
Series	1500								
Length	34"								
Steady Bearing Housing Assembly	3 -- 0970	Shock Absor. Appl.	28 UG 0200 - 0300	COWL. ASSY.	15 -- 0100	Castor			
Main Trans. Driveshaft Flange	3 AH 3201	STEER. GR. COMPL.	10 UK 0770	COWL APPLIC.	15 -- 02350	PAINT	#475- 6 199 Primer		
TRANSFER CASE - Front Upper Driveshaft Flange	3 DA 3201	Steer. Gear	10 UK 01200	CAB	15 UBK 0500	Shop List			
TRANSFER CASE - Rear Lower Driveshaft Flange	3 DC 3201	Pitman Arm	10 UK 01370	" Application	15 UBK 0800	POWER CARD			
Rear Axle Driveshaft Flange	3 RG 3201	Hght. Abv. Frame (T70) Series	38-1/8	Floor Board Installation	16 UGB 01700	Refer to:—( )			
TRANSFER CASE - Front Lower Driveshaft Flange	3 BL 3201A	Drag Link	10 UG 010	OILERS	Alemite				
Front Axle Driveshaft Flange	3 AM 3201								

Customer	U.S.A. Holabird Q.M. Depot	Model	U-2044	Wheelbase	128"	Frame Length	119"
Branch	Baltimore	Date Issued		Date Built		Date Shipped	
Chassis No.	U-2044-	Motor No.	(JXD)	Trans. Ser. No.		Cab No.		Sales Order No.	

CAR BUILDING RECORD

THE AUTOCAR COMPANY
(*) Special Items

ARDMORE, PA.

Hercules	4"	REAR AXLE	4 TLG	0270	Brake Disc	3 D	0381	Battery	16 ZRM 0440
ENGINE (JXD)	2 UG 080	5002-TW-X-1 Ratio 6.62	Brakes 16x3-1/2		Disc Brake & ppl.	10 HA	0990	Location Install.	In Cab --
Flywheel With Engine	006	Gear Carrier	4 BA	0170	FRAME	12 UK	710 A	Tool Box | Location Installation	In Cab 16 --
Pilot Brg.	3 A 0243	Brake Lining		052	Frame (Frt. End) Assembly 12	--	No. 1	HOOD	16 UBK 0950
Governor	2 UK 0650	Rr. Hub & Dr. Assem.	4 BP	0809	Frame (Center) Assembly 12	--	No. 2	HEADLIGHTS	16 UBK 0460
Oil Filter	2 UBA 0808	Rr. Wheel Adapter		0559	Frame (Rear End) Assembly 12	--	No. 3	Brakes	Air
Clutch	8 RL 030 A	Rear Hub	4 BP	004	Undrilled Frame Side Rails	12 UK	0112 0113	Brake Sketch	2308
Mfg. No. Long	12 CBC	Rr. HubCap Adptr.			Brake CrossShaft Cross Member	12 UG	4480	Trailer Conn.	On Brake Sketch
Carburetor Fuel Pump	6UG0510 UG 050	Rear Brake Drum	4 BP	021	SPRINGS, FRT.	13 UG	009	Hand Brake	H-15
Air Cleaner United Oil Bath ZBM	0330	Wheel Studs, R.H.		0489	Frt. Spring Clip	13UGA042 13 UK	042	Air Gauge Install.	25UBM01010
Air Cleaner Installation	6 UBK 0440	" " L.H.		0491	" " Shim	13 UG	014	WHEELS, Frt. Size	20 x 8 L Rim
Dist. & Gear	8 UG 0470	Drum Bolts			Thick End	Forward		" Part No.	9BJ0403
Mfg. No. DR-EX- 37343		Axle & Flng. Assy.		020	Frt. Spr. Keeper	13 UG	043	" Mfg. No.	33287
Coil 6-V.	8 SA 0120 A	Brake Type	Air		SPRS., REAR	13 UG	011	WHEELS, Rr. Size	20 x 8 L Rim
Mfg. No.	DR-528C	RADIATOR	5 UK	590	Rear Spr. Clip	13 SDS	041	" Part No.	9BJ 0403
Generator	16 BLF 0410	Fan (21 ")	5 UK	0120	" " Keeper	13 UG	092	" Mfg. No.	33287
Mfg. No.	DR-1102404	6-Blade Fan Belt	5 UK	055	Rr. Spg. Clip Plt.	13 UG	0111	Spare Wheel & Tire	9BJ0403
Voltage Regulator	16 BLG 0620	GASO. TANK	6 URK	4720	" " Shim	13 UG	0113-14	TIRES, Front	9.00-20
Mfg. No.	DR-5540	Location	LH Rail		Thick End	Rear		" Make	--
Starter	16 BLA 0560	Capacity	40 gallons		SPRGS., AUX.	13 UG	0156	TIRES, Rear	9.00-20
Mfg. No.	DR-720T	Aux. Gaso. Tank	6 --		Brkt. Clearance	1-1/2	"	" Make	--

DIRECTORY OF AUTOCAR FACTORY BRANCHES

THE AUTOCAR COMPANY   **ARDMORE, PA.**

All Autocar Factory Branches are Operated as The Autocar Sales & Service Co.
In This Listing "B" Indicates a Branch, "D" Indicates a Dealer and "S" Indicates a Service Station.*

STATE	*	CITY	ADDRESS	NAME	BRANCH MANAGER	DAY PHONE	NIGHT PHONE
California	S	Bakersfield	2211 Chester Avenue	Meagher-Morris Co., Inc.	C. G. Lee	Bakersfield 2211	Bakersfield 4863
	B	Fresno	1600 H Street	The Autocar S. & S. Co.	C. H. Vernon, V. P.	Fresno 2-0812	3-9015
	B	Los Angeles	1801 South Main Street	"	J. H. Phillips	PRospect 5421	Budlong 8-5587
	B	Oakland	4th and Alice Streets	"		Templebar 7115	Sweetwood 2394
	S	Sacramento	608 L Street	F. J. Coyle		Main 829	Main 8998-J
	B	San Diego	741 Front Street	The Autocar S. & S. Co.		Franklin 3451	Jackson 6228
	B	San Francisco	Bay Sh. Blvd. & San Bruno Ave.	"	J. H. Phillips	Valencia 0502	Valencia 5420
	S	Stockton	16 North Wilson Way	A. B. Coyle		Stockton 2-6516	Stockton 3-2485
Colorado	D	Denver	Stout at Broadway	Harrison Motors, Inc.		Keystone 2341	Spruce 3100 and Emerson 5117
Connecticut	B	Bridgeport	300 Post Rd. (Fairfield, P.O.)	The Autocar S. & S. Co.	G. G. Jones	Fairfield 9-0742	New Haven 8-0155
	S	Groton	169 Thames Street	Gott's Garage		Groton 3892	Groton 3892, 4237
	B	Hartford	2813 Main Street	The Autocar S. & S. Co.	H. F. Marnell	7-0451	
	B	New Haven	168 Columbus Avenue	"	C. D. Allen, D. M.	8-0155	For Night Service anywhere in Connecticut call New Haven 8-0155
	D	Norwich	199 W. Thames St.	Swanton & Co.		Norwich 3932	
	S	Stamford	60 Holly Place	Carlson's Garage		3-4755	
	D	Torrington	Railroad Square	E. J. Kelley Company		Torrington 9243	
Delaware	B	Wilmington	409 North Harrison Street	The Autocar S. & S. Co.	Howard Legg	Wilmington 21311	Wilmington 21633
Dist. of Col.	B	Washington	1073 Thirty-first St., N. W.	The Autocar S. & S. Co.	P. B. Lum	Michigan 4323	Columbia 4088-W Randolph 6858
Illinois	B	Chicago	1716 West Pershing Road	The Autocar S. & S. Co.	J. A. Donnelly, V. P.	Virginia 0210	same
Iowa	S	Des Moines	9th and Mulberry Streets	Beattie's Garage	Robert Beattie	4-3175	same
Kentucky	D	Louisville	215 South Second Street	Andriot Equipment Co.	E. N. Andriot	Wabash 4376	Wabash 1181
Maine	S	Portland	353 Cumberland Avenue	Madore's Auto Service		Portland 3-7481	Portland 2-3327
Maryland	B	Baltimore	27th and Sisson Streets	The Autocar S. & S. Co.	A. H. Bishop, V. P.	University 4300	Ches. 4611 Forrest 3151
	B	Salisbury	Willow and West Main Sts.	"	Jerry Valliant	Salisbury 1500	Salisbury 1123-M
Massachusetts	B	Boston	1168 Commonwealth Avenue	The Autocar S. & S. Co.	H. R. Gary, V. P.	Aspinwall 4450	same
	B	Fall River	93 President Avenue	"		2870	4926 & Dex 5300
	B	Lawrence	380 Andover Street	"		Lawrence 2-2135	Lawrence 2-1603
	B	New Bedford	So. Water cor. of School St.	"	O. D. Edwards	New Bedford 5143	New Bedf'd 5143 and Dexter 5300
	B	Springfield	1186 State Street	"	F. R. Holmes, Jr.	3-4140; 3-4149	4-3606
	B	Worcester	162 Shrewsbury Street	"		Worcester 3-6367	Worcester 3-6367
Michigan	B	Detroit	290 Piquette Avenue	The Autocar S. & S. Co.	B. F. Dunham, D.M	Madison 5285	same
Missouri	D	Kansas City	500 Pennway	Mason Truck & Trailer Co.	Tom Mason	Victor 1366	same
	B	St. Louis	2740 Locust Street	Autocar S. & S. Co. of Mo.	J. O. Warner, V. P.	Jefferson 0890	Goodfellow 2212
New Jersey	B	Camden	2041 Federal Street	The Autocar S. & S. Co.	John V. Myers	279	Camden 3348-R
	B	Jersey City	943 Communipaw Avenue	"		Bergen 3-7227	Mitchell 2-4040
	B	Newark	282 Jefferson Street	"	J. B. Rosenquest, D.M	Mitchell 2-4040	Mitchell 2-4040
	B	Paterson	740 Madison Avenue	"		Sherwood 2-2420	Mitchell 2-4040
	B	Trenton	131 Brunswick Avenue	"	John A. McCullough	Trenton 5663	Trenton 30494
New York	B	Albany	1030 Broadway	The Autocar S. & S. Co., Inc.	L. E. Harmon, D. M.	4-0105	Albany 5-4977 Albany 2-1985
	D	Auburn	5-7-9 Dill St. & 16 Water St.	Neese Bros. Garage	N. E. Nessel C. F. Burt	128	1937-R
	B	Binghamton	325 Clinton Street	Autocar S. & S. Co.	F. A. Reynolds	64531	2-8844
	B	Bronx	1170 Randall Ave.	The Autocar S. & S. Co., Inc.		Dayton 9-4460	Watkins 9-7900
	B	Brooklyn	930 Bedford Avenue	"		Evergreen 7-8200	Watkins 9-7900
	B	Buffalo	1122 Niagara Street	"	G. M. Wilkins, D.M.	Lincoln 4425	Triangle 7880 Parkside 1858
	D	Jamestown	108 Harrison Street	J. N. K. Machine Company		66135	
	D	Keeseville		A. Brelia		39-R	154-R
	S	Newburgh	170 S. Robinson Ave.	Truck Repair Service Co.	John H. Lewis Charles F. Booth	Newburgh 1188	same
	B	New York	555 West 23rd Street	The Autocar S. & S. Co., Inc.	F. D. Wait, V. P.	Watkins 9-7900	Watkins 9-7900
	S	Poughkeepsie	100 Church Street	Brigg's Garage		2917	1631-M
	B	Rochester	40 Mount Hope Avenue	The Autocar S. & S. Co., Inc.	Lloyd A. Wenrich	Stone 4110	Stone 2529-L
	B	Syracuse	730-32 East Water Street	"	M. K. Thomson	Syracuse 6-1509	3-0568 3-3345
	D	Utica	324 Lafayette Street	Harry Heiman, Inc.		4-1127	
	D	Westfield	Peck Motor Co.			82-J	
North Carolina	B	Charlotte	Cor. E. Trade & McDowell Sts	The Autocar S. & S. Co.	J. H. Jenkins	3-3250	2-3766
Ohio	S	Akron	1520 E. Market St.	Earle Mansfield & Co., Inc.		Jefferson 8155	same
	S	Canton	343 Eighteenth St., N. W.	M. J. Scanlon		28237	same
	B	Cincinnati	243 W. McMicken Avenue	The Autocar S. & S. Co.	Carl B. Fehr	Main 4794	Melrose 1176
	B	Cleveland	1961 East 61st Street	"	R. D. Peo, V.P.	H'ndrs'n 7970-1-2	same
	B	Columbus	582 West Rich Street	"	M. H. Lenhart	Main 5271-5272	University 2200
	B	Youngstown	630 Marshall Street	"	E. E. Dickason	76124	76124
Pennsylvania	B	Bryn Mawr	736 Lancaster Avenue	The Autocar S. & S. Co.		Bryn Mawr 2650	Ardmore 3582-W
	B	Chester	5th and Barclay Streets	"	R. Laughead	Chester 8015	Chester 7534
	S	Harrisburg	Camp Hill	L. B. Smith, Inc.		Harrisburg 7331	same
	S	Lancaster	Manheim Pike	V. H. Knowles		Lancaster 2-7786	
	B	Philadelphia	34th St. & Indiana Ave.	The Autocar S. & S. Co.	J. E. Higgins, V. P.	Sagamore 4010	same
	B	Pittsburgh	Baum Blvd. at Liberty Ave.	"	H. L. White	Schenley 8383	Montrose 8148
	B	Reading	133-35 Chestnut Street	"	E. M. Shields	4-5771	9-1150
	D	Scranton	620 West Linden Street	Ted V. Rodgers Company		Scranton 2-3015	Scranton 8581
Rhode Island	D	Newport	Long Wharf	Perry Garage		Newport 260	
	B	Providence	Branch Ave. at N. Main St.	The Autocar S. & S. Co.	W. J. Fowler D. M.	Dexter 5300	same
Texas	D	Dallas	2311 Main Street	Boling-Duggan	King Duggan	7-9231	same
	D	Houston	Rusk & LaBranch Sts.	Eller & Gripp Company	G. W. Howell—Parts	Fairfax 5321	Hadley 3084
Virginia	B	Norfolk	Granby Street at 27th	The Autocar S. & S. Co.	P. R. Wood	27654	2-7897
	B	Richmond	2804 West Broad Street	"	C. M. Hogarth D.M.	5-1787	2-5003
	B	Roanoke	4th St. and Shenandoah Ave.	"	J. L. Knott	3-0904	2-8168
Wisconsin	S	Milwaukee	1556 North Farwell Avenue	Larry Humphrey		Broadway 9707	same
Hawaii	D	Honolulu	P. O. Box 78	Grace Brothers, Ltd.			
Dominion of Canada Province of Ontario	D	Toronto	40 Oxford Street	Elliott Autocar Trucks	J. E. Elliott	Midway 7755	Hudson 2648

May 1, 1940   PRINTED IN U. S. A.

DIRECTORY OF
OFFICERS AND DEPARTMENT HEADS

The Autocar Company, Ardmore, Pa.

Robert P. Page, Jr., *President*

John C. Taney, *Vice President & Treasurer*
B. B. Bachman, *Vice President in charge of Engineering*
C. A. Borton, *Vice President in charge of Manufacturing*
W. H. Brearley, *Secretary & Assistant Treasurer (Insurance, Taxes & Legal Matters*
H. M. Coale, *Vice President in charge of Sales*
C. R. C. Custer, *Vice President & Comptroller*

James Battersby, *Assistant Sales Mgr.*
John E. Bower, *Purchasing Agent*
Edward F. Coogan, *Sales Manager*
W. J. Diederichs, *Metallurgist*
Adolf Gelpke, *Chief Engineer*

F. C. Hubley, *General Service Manager*
F. A. McMonigle, *Auditor*
E. M. Minshall, *Service Department*
W. J. Purvis, *Treasury Department, (Credits & Collections)*

A. A. Reiter, *Office Manager & Factory Accountant*
E. D. Sirrine, *Transportation Engineer*
Robert F. Wood, *Advertising Manager*

The Autocar Sales & Service Company (New Jersey)

Robert P. Page, Jr., *President*
James Battersby, *Vice President*
Alexander H. Bishop, *Vice President*
W. H. Brearley, *Secy. & Asst. Treas.*
H. M. Coale, *Vice President*
Edward F. Coogan, *Vice President*

C. R. C. Custer, *Treas. & Asst. Secy.*
Charles E. Doling, *Vice President*
Joseph A. Donnelly, *Vice President*
C. Eustace Dwyer, *Vice President*
H. R. Gary, *Vice President*
James E. Higgins, *Vice President*

C. F. Iredell, *Asst. Secy.*
Ray D. Peo, *Vice President*
John C. Taney, *Vice President*
C. H. Vernon, *Vice President*
F. D. Wait, *Vice President*

Autocar Sales & Service Company of Missouri

Robert P. Page, Jr., *President*
W. H. Brearley, *Secretary*

H. M. Coale, *Vice President*
C. R. C. Custer, *Treasurer*

John C. Taney, *Vice President*
J. O. Warner, *Vice President*

Autocar Sales & Service Company of Texas

Robert P. Page, Jr., *President*
James Battersby, *Vice President*

W. H. Brearley, *Secretary*
H. M. Coale, *Vice President*

John C. Taney, *Treasurer*

DISTRICT AND TERRITORIAL GROUPINGS
OF THE AUTOCAR SYSTEM OF BRANCHES, DEALERS AND SERVICE STATIONS

DISTRICT	DISTRICT HEADQUARTERS	DISTRICT MANAGER	DISTRICT BRANCHES AND SERVICE STATIONS
Boston	Boston, Mass.	H. R. Gary, V.P.	Boston, Lawrence and Worcester, Mass. Portland, Me.
Providence	Providence, R. I.	W. J. Fowler	Providence and Newport, R. I., New Bedford and Fall River, Mass., Groton and Norwich, Connecticut
New Haven	New Haven, Conn.	C. D. Allen	Springfield, Mass. Bridgeport, Hartford, New Haven, Stamford and Torrington, Conn.
Metropolitan New York	New York, N. Y.	Frank D. Wait, V.P.	New York City, Brooklyn and Bronx, N. Y.
Albany	Albany, N. Y.	L. E. Harmon	Albany, Keesville and Poughkeepsie, N. Y.
Buffalo	Buffalo, N. Y.	Geo. M. Wilkins	Buffalo, Auburn, Binghamton, Jamestown, Syracuse, Rochester and Utica, N. Y.
Newark	Newark	J. B. Rosenquest	Newark, Jersey City, Paterson, N. J.
Philadelphia	Philadelphia, Pa.	James E. Higgins, V.P.	Philadelphia, Bryn Mawr, Chester, Harrisburg, Lancaster, Reading, Scranton, Pa. Wilmington, Del. Camden and Trenton, N. J.
Baltimore	Baltimore, Md.	A. H. Bishop, V.P.	Baltimore, Md. and Washington, D.C.
Richmond	Richmond, Va.	C. M. Hogarth,	Salisbury, Md. Richmond, Norfolk and Roanoke, Va.
Southern	Charlotte, N. C.	J. H. Jenkins	Atlantic Seaboard States South of Virginia
Pittsburgh	Pittsburgh, Pa.	H. L. White	Western Pennsylvania
Cleveland	Cleveland, Ohio	Ray D. Peo, V.P.	Cleveland, Columbus, Cincinnati, Canton, Youngstown and Akron, Ohio. Louisville, Ky.,
Michigan	Detroit	B. F. Dunham	State of Michigan
Chicago	Chicago, Ill.	J. A. Donnelly, V.P.	Chicago, Ill., Des Moines, Iowa, and Milwaukee, Wis.
South Western	St. Louis, Mo.	J. O. Warner, V.P.	St. Louis and Kansas City, Mo., Tulsa, Oklahoma. Dallas and Houston, Texas
Pacific	Los Angeles, Cal.	C. H. Vernon, V.P.	Los Angeles, San Francisco, Bakersfield, Sacramento, Stockton, Fresno, San Diego and Oakland, Cal. Denver, Colorado.

Mfg. No.	DR-720T	Aux. Gaso. Tank	6 --	Brkt. Clearance	1-1/2 "	" Make	16 omit
MAIN TRANS.	8 UK 6530	Aux. Gaso. T'k Ins.	6 --	Aux. Spr. Brkts.	13 DK 0154	Tire Carrier	12UK402
Mfg. No. Clark	230F	Location	-- --	Run. Brd. & Spl. Grd. & Fdr. Assem. R.H.	14 -- 070	BUMPER	
TRANSFER CASE	19 BK 0440	MUFFLER	7 TE 010	Run. Brd. & Spl. Grd. & Fdr. Assem. L.H.	14 -- 080	START. CRANK	1 ZN 010
TIMKEN	T-2-B-7-4	Exhaust Pipe	7 UK 003	FENDER, R.H.	14 URK 052 A	Grille Radiator Guard	5URK1540 0140
Speedo Cable SK X Extension	16 LE 0800	Exhaust Pipe Ext.	7 -- 040	" L.H.	14 URK 053 A	Wheel Brg. Wrenches	Fr.-- None Rr.-- None
" Adapter	SW 1.071	FRONT AXLE	9 BKW 6430	Spl. Guard, R.H.	14 URK 021	Wheel Wrenches	16SA0743-44
" Gear	15T-205052	Ratio 6.62:1	Brakes 16x3-1/2 With Axle	" " L.H.	14 URK 022	Ignition Key No.	
Front Drive Shaft	30 LD 4110	Steering Arm		Runn. Bd.-R.H.	14 URK 343	Cab Door Key No.	
Series 1500	Length 11-1/8	Fr. Hub & Dr. Assem.	9 BP 0809	Runn. Bd.-L.H.	14 URK 344	Tool Box Key No.	
Front Axle Driveshaft	30 ZC 5330B	Frt. Hub	9 BP 004	" " Iron, R.H.	14 -- 001	Chassis Wght, Frt.	
Series 1410	Length 33"	Frt. Hub Cap Adapter	9 --	" " " L.H.	14 --	" " Rr.	
Rear Driveshaft	30 DD 5340	Frt. Brake Drum	9 BP 0232	" " Brace (F)	14 -- 078	Tot. Chassis Wght.	
Series 1500	Length 34"	Brake Lining	9 --	" " " (R)	14 -- 078	Wheel Cuts	R.H. : L.H. :
Steady Bearing Housing Assembly	3 -- 0970	Brake Type	Air	COWL ASSY.	15 -- 0100	Castor	
Main Trans. Driveshaft Flange	3 AH 3201	Shock Absor. Appl.	28 UG 0200 - 0300	COWL APPLIC.	15 -- 02350	PAINT #475- 6199 Primer	
TRANSFER CASE - Front upper Driveshaft Flange	3 DA 3201	STEER. GR. COMPL.	10 UK 0770	CAB	15 UBK 0500	Shop List	
TRANSFER CASE - Rear Lower Driveshaft Flange	3 DC 3201	Steer. Gear	10 UK 01200	" Application	15 UBK 0800	POWER CARD	
Rear Axle Driveshaft Flange	3 RG 3201	Pitman Arm	10 UK 01370	Floor Board Installation	16 UGB 01700		
TRANSFER CASE - Front lower Driveshaft Flange	3 BL 3201A	Hght.Abv.Frame (T70) Series	38-1/8	OILERS	Alemite	Refer to:--( )	
Front Axle Driveshaft Flange	3 AM 3201	Drag Link	10 UG 010				

Customer	U.S.A. Holabird Q.M. Depot		Model U-2044		Wheelbase 128"	Frame Length 119"
Branch	Baltimore		Date Issued		Date Built	Date Shipped
Chassis No. U-2044-		Motor No.	(JXD)	Trans. Ser. No.	Cab No.	Sales Order No.

INSTRUCTION BOOK AND PARTS LIST

INDEX

U-2044 - U.S.A. 1940

Bulletin Number	Parts List Number	Contents	Page Number
-	-	Directory	
-	-	Car Building Record	
460	-	General Operation	1
-	-	Lubrication	2
585	772	Engine	3
689	-	Engine Gasket Kit	11
599	689	Oil Filter	12
700	866	Governor	14
633	645	Clutch	18
682	830	Main Transmission	24
697	860	Transfer Case	32
-	457X	Driveshaft Assembly	37
590	839	Front Driveshaft	38
397	478	Intermediate & Rear Driveshaft	40
397	464	Rear Driveshaft	43
703	870	Driveshaft Brake	44
695	847	Rear Axle	47
315	-	Adjustment of Spiral Bevel Gears	54
-	871	Rear Brake	56
-	864	Rear Shock Absorber	57
-	855	Engine Cooling System	58
623	534	Water Pump	59
663	836	Fan & Fan Drive	61
-	853	Fuel System	63
691	869	Carburetor	64
594	862	Fuel Pump	69
688	-	Air Cleaner	71
-	852	Exhaust System	72
702	867	Ignition & Distributor	73
696	848	Front Driving Axle	77
-	872	Front Brake	85
-	863	Front Shock Absorber	86
685	827	Steering Gear	87
-	865	Drag Link	92
-	845	Frame & Brackets	93
681	829	Front Spring Mounting	94
-	825	Front Spring	96
692	837	Rear Spring Mounting	97
-	826	Rear Spring	99
-	824	Rear Auxiliary Spring	100
-	842	Cab	101
-	849	Fenders and Sheet Metal	108
693	-	Wiring Diagram	109
-	843	Electrical Units and Wiring	111
-	843 (Supp.)	Generator	114
-	843 (Supp.)	Starting Motor	116
#2308	858	Air Brake Hook-Up Diagram	118
313	-	Air Brake (Also B.W. Folder)	121
400	857	Wheels & Bearings	128

SERVICE BULLETIN No. 460

THE AUTOCAR COMPANY, ARDMORE, PENNSYLVANIA
SERVICE DEPARTMENT

SUBJECT   GENERAL OPERATION   ALL MODELS

It is the aim of THE AUTOCAR COMPANY to deliver to the customer a high class commercial vehicle, properly adjusted and in perfect running condition to effectively operate in the service for which it was purchased.

Periodic inspection, with prompt attention to loose connections, adjustments, lubrication and minor repairs will enable the owner to fully develop the usefulness of this vehicle at the lowest possible operating cost.

Special attention should be given to lubrication as regular attention to this detail will have the greatest effect in keeping down maintenance costs.

In general, the following practice is recommended.

ENGINE

Change the oil in the crankcase at regular intervals and use a well refined oil of proper grade. Grade (Viscosity) is very important as too light or too heavy an oil may damage the engine. FOR EXACT RECOMMENDATIONS ON ENGINE OIL AND OTHER LUBRICANTS, REFER TO BULLETIN ON LUBRICATION.

BRAKES

Hand and foot (service) brakes should be inspected periodically for adjustment. Adjust to maintain reserve travel at all times so that pedal will not strike floorboards even under most extreme braking conditions. FOR DETAILS ON BRAKE ASSEMBLY, ADJUSTMENTS AND LUBRICATION, REFER TO BULLETINS ON BRAKES.

COASTING

Free coasting should be discouraged. WHEN COASTING, THE CAR SHOULD BE IN GEAR AND THE CLUTCH ENGAGED. Remember that a loaded truck weighs many tons and for safety sake, descend long, heavy grades in gear.

IGNITION

Distributor breaker points should be kept clean and adjusted to approximately .018-.024 maximum opening in order to maintain uniform spark action. Spark plug points are set at .022-.025 and, as this gap will increase in service, the points should be occasionally checked and reset. Replacements should be made with Champion plugs of the same type as original equipment.

FUEL SUPPLY SYSTEM

Clean sediment cup under fuel pump at regular intervals. Excessive sediment or water in the fuel supply system will cause trouble.

AIR CLEANER

Where dust conditions are very severe as in road or quarry operation, in excavating work, sand pits and around cement mills, it is necessary to clean and re-oil the cleaner DAILY. Under ordinary conditions the cleaner should be cleaned WEEKLY.

STEERING

Free steering without excessive play is essential to safe operation. Means are provided in the steering gear to take up play and one grease fitting supplies lubricant to the working parts. The connections between the steering gear and wheels are all thoroughly bushed and each bearing is provided with a fitting for lubrication. Inspect each connection and re-bush when necessary. Lubricate at weekly intervals.

SPRING CLIPS

Tighten spring clips at occasional intervals, using the extra heavy wrenches provided in the tool kit for this purpose. A loose spring clip may permit either axle to shift out of alignment, causing spring plate breakage, hard steering, excessive tire wear or poor brake action.

LUBRICATION CHART AUTOCAR MOTOR TRUCKS

THE AUTOCAR COMPANY, ARDMORE, PA.

MODEL U-2044

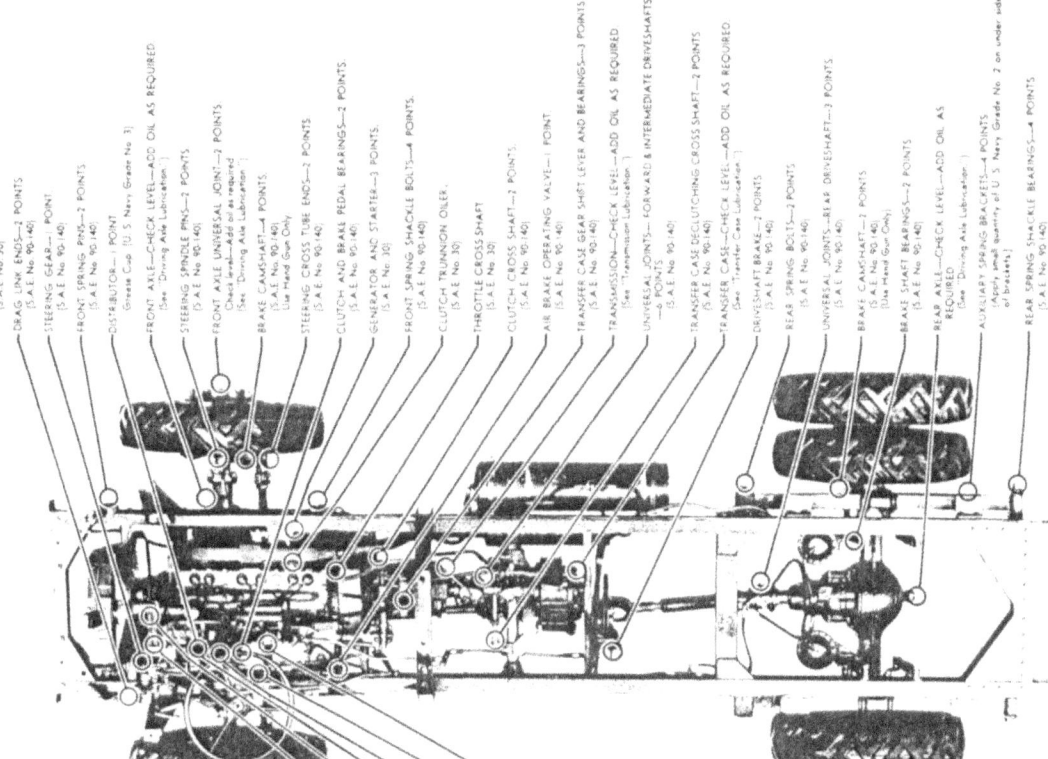

WEEKLY or EVERY 1000 MILES

When two grades of S.A.E. are specified—The first is for Winter—The second for Summer.

- ALL DOOR LOCKS AND HINGES (S.A.E. No. 30)
- DRAG LINK ENDS—2 POINTS (S.A.E. No. 90-140)
- STEERING GEAR—1 POINT (S.A.E. No. 90-140)
- FRONT SPRING PINS—2 POINTS (S.A.E. No. 90-140)
- DISTRIBUTOR—1 POINT (Grease Cup (U.S. Navy Grade No. 3))
- FRONT AXLE—CHECK LEVEL—ADD OIL AS REQUIRED (See Driving Axle Lubrication)
- STEERING SPINDLE PINS—2 POINTS (S.A.E. No. 90-140)
- FRONT AXLE UNIVERSAL JOINT—2 POINTS (S.A.E. No. 90-140) (Check oil level—Add oil as required. See Driving Axle Lubrication)
- BRAKE CAMSHAFT—4 POINTS (S.A.E. No. 90-140) (Use Hand Gun Only)
- STEERING CROSS TUBE ENDS—2 POINTS (S.A.E. No. 90-140)
- CLUTCH AND BRAKE PEDAL BEARINGS—2 POINTS (S.A.E. No. 90-140)
- GENERATOR AND STARTER—2 POINTS (S.A.E. No. 30)
- FRONT SPRING SHACKLE BOLTS—4 POINTS (S.A.E. No. 90-140)
- CLUTCH TRUNNION OILER (S.A.E. No. 30)
- THROTTLE CROSS SHAFT
- CLUTCH CROSS SHAFT—2 POINTS (S.A.E. No. 90-140)
- AIR BRAKE OPERATING VALVE—1 POINT (S.A.E. No. 90-140)
- TRANSFER CASE GEAR SHIFT LEVER AND BEARINGS—3 POINTS (S.A.E. No. 90-140)
- TRANSMISSION—CHECK LEVEL—ADD OIL AS REQUIRED (See Transmission Lubrication)
- UNIVERSAL JOINTS—FORWARD & INTERMEDIATE DRIVESHAFTS—6 POINTS (S.A.E. No. 90-140)
- TRANSFER CASE DECLUTCHING CROSS SHAFT—2 POINTS (S.A.E. No. 90-140)
- TRANSFER CASE—CHECK LEVEL—ADD OIL AS REQUIRED (See Transfer Case Lubrication)
- DRIVESHAFT BRAKE—2 POINTS (S.A.E. No. 90-140)
- REAR SPRING BOLTS—2 POINTS (S.A.E. No. 90-140)
- UNIVERSAL JOINTS—REAR DRIVESHAFT—3 POINTS (S.A.E. No. 90-140)
- BRAKE CAMSHAFT—2 POINTS (S.A.E. No. 90-140) (Use Hand Gun Only)
- BRAKE SHAFT BEARINGS—2 POINTS (S.A.E. No. 90-140)
- REAR AXLE—CHECK LEVEL—ADD OIL AS REQUIRED (See Driving Axle Lubrication)
- AUXILIARY SPRING BRACKETS—4 POINTS (Apply small quantity of U.S. Navy Grade No. 2 on under side of brackets.)
- REAR SPRING SHACKLE BEARINGS—4 POINTS (S.A.E. No. 90-140)

ENGINE LUBRICATION

OIL SPECIFICATIONS
FOR SUMMER—Use S.A.E. No. 30 OIL
FOR WINTER—Use S.A.E. No. 20W OIL
FOR TEMPERATURES BELOW 10 FAHRENHEIT OR FOR INTERMITTENT WINTER SERVICE—Use S.A.E. No. 20 OIL

CHANGE OF OIL
In severe service change oil every 1000 miles. Under normal working conditions change oil every 2500 miles. Crankcase capacity is 6 quarts.

CLEAN OIL PAN—OIL SCREEN
In the spring and fall when changing grade of oil remove and scrub out oil pan and crankcase also remove and wash oil screen.

When replacing screen be sure to place the felt washer in position at bottom of screen.

OIL PRESSURE
Warm the engine, warm oil pressure gauge on the instrument panel should show 35 pounds pressure at governed speed and 5 pounds when idling.

Oil pressure is controlled by a valve on the oil pump which can be adjusted through a pipe plug opening on the left side of the oil pan using special wrenches provided for this purpose.

As bearings become worn more oil will escape at these points resulting in some drop in pressure. Do not readjust this pressure until plays taken out of bearings at otherwise overoiling will result.

OIL FILTER
The filter cartridge should be replaced at periods of 5,000 to 10,000 miles, depending upon the nature of service. The cartridge can be replaced at very low cost. DO NOT ATTEMPT TO CLEAN AND REUSE AN OLD CARTRIDGE. It is loaded with dirt and other foreign substances some of which cannot be dissolved in gasoline. The cartridge should be replaced when oil begins to darken on bayonet gauge.

To remove the cartridge unscrew the handle on the top of the filter unit, release the lock bar from the upright the rod, and lift out.

When replacing cartridge, see that gasket seats and ring groove are clean.

Refill crankcase to proper level plus one quart for the filter. Run motor and check for leaks.

DAILY or EVERY 250 MILES

EXAMINE ENGINE CRANKCASE OIL GAUGE. KEEP OIL LEVEL TO 4/4 MARK. DON'T OVERFILL OR ALLOW LEVEL TO DROP BELOW 3/4 MARK.

CHANGE OIL EVERY 2500 MILES (See Engine Lubrication)

EVERY 5000 MILES

- FRONT WHEEL BEARINGS—2 POINTS. Remove wheels and repack with U.S. Navy Grade No. 2 Every 5000 miles or 6 months.
- ENGINE TRUNNION BRACKET—1 POINT (S.A.E. No. 30)
- FAN SHAFT BEARINGS—1 POINT (S.A.E. No. 90-140)
- DISTRIBUTOR ROTOR SHAFT—1 POINT (Few drops of S.A.E. No. 10 Engine Oil on felt wick under rotor.)
- WATER PUMP—1 POINT (Grease Cup U.S. Navy Waterproof Lubricant S.A.E. No. 90-100)
- SPARK AND THROTTLE CONTROLS—ALL JOINTS (S.A.E. No. 30)
- OIL FILTER—CHANGE AS INSTRUCTED (See instructions at left)
- REAR WHEEL BEARINGS—2 POINTS. Remove wheels and repack with U.S. Navy Grade No. 2— Every 5000 miles or 6 months.

TRANSMISSION, TRANSFER CASE AND DRIVING AXLE LUBRICATION

Change oil in Transmission and Rear Axle every three months or every 10,000 miles. Frequent changing is particularly important when using E.P. (extreme pressure) lubricants, since these oils oxidize and increase in viscosity more rapidly than ordinary oils.

Units should be thoroughly drained and flushed before refilling.

Use the following specifications:

For Summer—S.A.E. No. 140 oil. Must not channel at 35 F.

For Winter—S.A.E. No. 90 oil. Must not channel at 10 F.

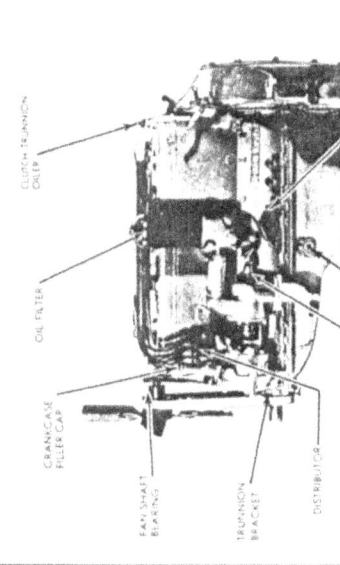

Labels: CLUTCH TRUNNION OILER; OIL LEVEL BAYONET GAUGE; OIL PRESSURE ADJUSTMENT; OIL FILTER; CRANKCASE FILLER CAP; WATER PUMP GREASE CUP; DISTRIBUTOR; TRUNNION BRACKET; FAN SHAFT BEARING

PROPER LUBRICATION IS THE MOST IMPORTANT THING IN THE CARE OF THE MOTOR TRUCK

LUBRICATION CHART

THE AUTOCA
MO

ENGINE LUBRICATION

OIL SPECIFICATIONS

FOR SUMMER—Use S.A.E. No. 30 OIL

FOR WINTER—Use S.A.E. No. 20W OIL

FOR TEMPERATURES BELOW 10 FAHRENHEIT OR FOR INTERMITTENT WINTER SERVICE—Use S.A.E. No. 20 OIL

CHANGE OF OIL

In severe service change oil every 1000 miles. Under normal working conditions change oil every 2500 miles. Crankcase capacity is 6 quarts.

CLEAN OIL PAN—OIL SCREEN

In the spring and fall when changing grade of oil, remove and scrub out oil pan and crankcase, also remove and wash oil screen.

When replacing screen, be sure to place the felt washer in position at bottom of screen.

OIL PRESSURE

With the engine warm, oil pressure gauge on the instrument panel should show 35 pounds pressure at governed speed and 5 pounds when idling.

Oil pressure is controlled by a valve on the oil pump which can be adjusted through a pipe plug opening on the left side of the oil pan using special wrenches provided for this purpose.

As bearings become worn more oil will escape at these points resulting in some drop in pressure. Do not readjust this pressure until play is taken out of bearings as otherwise overoiling will result.

OIL FILTER

The filter cartridge, should be replaced at periods of 5,000 to 10,000 miles, depending upon the nature of service. The cartridge can be replaced at very low cost. DO NOT ATTEMPT TO CLEAN AND REUSE AN OLD CARTRIDGE. It is loaded with dirt and other foreign substances some of which cannot be dissolved in gasoline. The cartridge should be changed when oil begins to darken on bayonet gauge.

To remove the cartridge, unscrew the handle on the top of the filter unit, release the lock bar from the upright tie rods and lift out.

When replacing cartridge, see that gasket seats and ring groove are clean.

Refill crankcase to proper level plus one quart for the filter. Run motor and check for leaks.

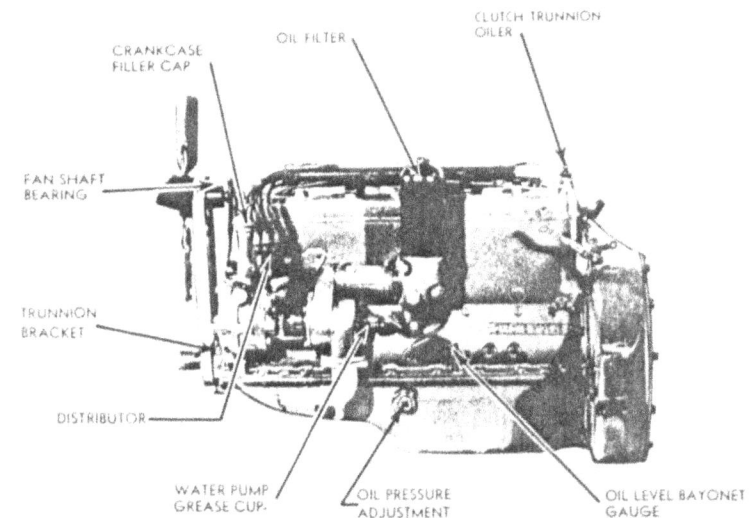

PROPER LUBRICATION IS THE M

AUTOCAR MOTOR TRUCKS
COMPANY, ARDMORE, PA.
MODEL U-2044

DAILY or EVERY 250 MILES

EXAMINE ENGINE CRANKCASE OIL GAUGE. KEEP OIL LEVEL TO 4/4 MARK. DON'T OVERFILL OR ALLOW LEVEL TO DROP BELOW 2/4 MARK.

CHANGE OIL EVERY 2500 MILES
(See "Engine Lubrication.")

EVERY 5000 MILES

FRONT WHEEL BEARINGS—2 POINTS
Remove wheels and repack with U. S. Navy Grade No. 2 Every 5000 miles or 6 months

ENGINE TRUNNION BRACKET—1 POINT
(S.A.E. No. 90-140)

FAN SHAFT BEARINGS—1 POINT
(S.A.E. No. 90-140)

DISTRIBUTOR ROTOR SHAFT—1 POINT
(Few drops of S.A.E. No. 30 Engine Oil on felt wick under rotor.)

WATER PUMP—1 POINT
Grease Cup (U. S. Navy Waterproof Lubricant S.A.E. No. 60-100)

SPARK AND THROTTLE CONTROLS—ALL JOINTS
(S.A.E. No. 30)

OIL FILTER—CHANGE AS INSTRUCTED
(See Instructions at Left.)

REAR WHEEL BEARINGS—2 POINTS
Remove wheels and repack with U. S. Navy Grade No. 2 Every 5000 miles or 6 months

TRANSMISSION, TRANSFER CASE AND DRIVING AXLE LUBRICATION

Change oil in Transmission and Rear Axle every three months or every 10,000 miles. Frequent changing is particularly important when using E. P. (extreme pressure) lubricants, since these oils oxidize and increase in viscosity more rapidly than ordinary oils.

Units should be thoroughly drained and flushed before refilling.

Use the following specifications:

For Summer—S.A.E. No. 140 oil. Must not channel at 35°F.

For Winter—S.A.E. No. 90 oil. Must not channel at 0°F.

WEEKLY or EVERY 1000 MILES

When two grades of S.A.E. are specified—The first is for Winter— The second for Summer.

ALL DOOR LOCKS AND HINGES
(S.A.E. No. 30)

DRAG LINK ENDS—2 POINTS
(S.A.E. No. 90-140)

STEERING GEAR—1 POINT
(S.A.E. No. 90-140)

FRONT SPRING PINS—2 POINTS
(S.A.E. No. 90-140)

DISTRIBUTOR—1 POINT
Grease Cup (U. S. Navy Grade No. 3)

FRONT AXLE—CHECK LEVEL—ADD OIL AS REQUIRED.
(See "Driving Axle Lubrication.")

STEERING SPINDLE PINS—2 POINTS
(S.A.E. No. 90-140)

FRONT AXLE UNIVERSAL JOINT—2 POINTS
Check level—Add oil as required.
(See "Driving Axle Lubrication.")

BRAKE CAMSHAFT—4 POINTS
(S.A.E. No. 90-140)
Use Hand Gun Only

STEERING CROSS TUBE ENDS—2 POINTS
(S.A.E. No. 90-140)

CLUTCH AND BRAKE PEDAL BEARINGS—2 POINTS
(S.A.E. No. 90-140)

GENERATOR AND STARTER—3 POINTS
(S.A.E. No. 30)

FRONT SPRING SHACKLE BOLTS—4 POINTS
(S.A.E. No. 90-140)

CLUTCH TRUNNION OILER
(S.A.E. No. 30)

THROTTLE CROSS SHAFT
(S.A.E. No. 30)

CLUTCH CROSS SHAFT—2 POINTS
(S.A.E. No. 90-140)

AIR BRAKE OPERATING VALVE—1 POINT
(S.A.E. No. 90-140)

TRANSFER CASE GEAR SHIFT LEVER AND BEARINGS—3 POINTS
(S.A.E. No. 90-140)

TRANSMISSION—CHECK LEVEL—ADD OIL AS REQUIRED
(See "Transmission Lubrication.")

UNIVERSAL JOINTS—FORWARD & INTERMEDIATE DRIVESHAFTS—6 POINTS
(S.A.E. No. 90-140)

TRANSFER CASE DECLUTCHING CROSS SHAFT—2 POINTS
(S.A.E. No. 90-140)

TRANSFER CASE—CHECK LEVEL—ADD OIL AS REQUIRED
(See "Transfer Case Lubrication.")

DRIVESHAFT BRAKE—2 POINTS
(S.A.E. No. 90-140)

REAR SPRING BOLTS—2 POINTS
(S.A.E. No. 90-140)

UNIVERSAL JOINTS—REAR DRIVESHAFT—3 POINTS
(S.A.E. No. 90-140)

BRAKE CAMSHAFT—2 POINTS
(S.A.E. No. 90-140)
(Use Hand Gun Only)

BRAKE SHAFT BEARINGS—2 POINTS
(S.A.E. No. 90-140)

REAR AXLE—CHECK LEVEL—ADD OIL AS REQUIRED
(See "Driving Axle Lubrication.")

AUXILIARY SPRING BRACKETS—4 POINTS
(Apply small quantity of U. S. Navy Grade No. 2 on under side of brackets.)

REAR SPRING SHACKLE BEARINGS—4 POINTS
(S.A.E. No. 90-140)

MOST IMPORTANT THING IN THE CARE OF THE MOTOR TRUCK

SERVICE BULLETIN   No. 585

THE AUTOCAR COMPANY, ARDMORE, PENNSYLVANIA
SERVICE DEPARTMENT

SUBJECT   HERCULES JX ENGINES   -   FOR CAB OVER ENGINE MODELS

CONSTRUCTION

Parts on these engines are interchangeable except for those affected by the difference in bore.

These engines vary from the stock JX engines in the following detail: Aluminum pistons, Tocco hardened crankshaft, cadmium nickel thin shell rod and main bearings, inserted exhaust valve seats and chrome nickel iron block.

Compression ratio is 5.8 to 1 on both engines.

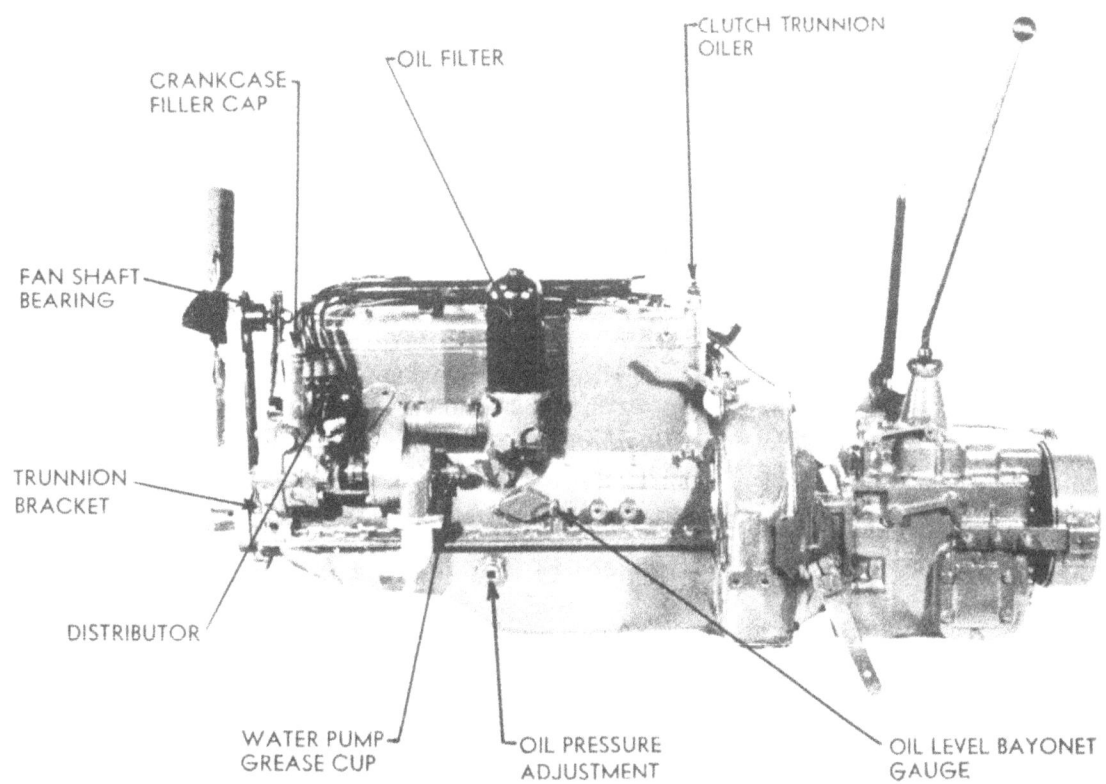

LUBRICATION

Oil Specifications - For summer, use S.A.E. No. 30 oil. For winter, use S.A.E. No. 20W oil. For temperatures below 10° F. or for intermittent winter service, use S.A.E. No. 20 oil.

Change of Oil - In severe service change oil every 1,000 miles. Under normal working conditions, change oil every 2,500 miles. Crankcase capacity is 6 quarts.

Clean Oil Pan - Oil Screen - In the spring and fall when changing grade of oil, remove and scrub out oil pan and crankcase. Also remove and wash oil screen. When replacing screen, be sure to place the felt washer in position at bottom of screen.

Oil Pressure - With the engine warm, oil pressure gauge on the instrument panel should show 35 pounds pressure at governed speed and 5 pounds when idling. Oil pressure is controlled by a valve on the oil pump which can be adjusted through a pipe plug opening on the left side of the oil pan using special wrenches provided for this purpose.

As bearings become worn, more oil will escape at these points resulting in some drop in pressure. Do not readjust this pressure until play is taken out of bearings as otherwise overoiling will result.

Oil Filter - The filter should be cleaned and repacked at 5,000 to 10,000 mile intervals depending upon the nature of service. Clean and repack when oil begins to darken on bayonet gauge.

To clean, remove plug at base and drain. Remove cap nut at top of shell, take off shell and lift out filter element. Remove wire ring and perforated washer at each end of element and pull out and discard filtering material.

Wash element in gasoline and repack as tightly as possible with a soft grade of white cotton waste free of lumps. When replacing shell, see that gasket seats and ring groove are clean.

Refill crankcase to proper level plus one quart for the filter. Run motor and check for leaks.

VALVE ADJUSTMENTS

Valve tappet clearance on both the intake and the exhaust valves is .006" adjusted when the engine is hot.

FLYWHEEL MARKINGS

Dead center is the only mark on the flywheel, placed on the front of the wheel where it can be observed through the hole in the flywheel housing back of the water pump.

Firing order is 1-5-3-6-2-4.

TIMING

The distributor timing on the Hercules JXB and JXC engines will be set to break at top dead center with full manual advance. The timing opening is a 7/8" hole located on the left hand side of the bell housing. Top dead center is indicated when the DC mark on the flywheel is lined up with the mark across the timing hole.

BEARING ADJUSTMENTS

Rod bearings are fitted with shims on one side only with a running clearance of .002".

Main bearings are fitted with shims on both sides with a running clearance of .0025" to .0035".

End thrust of the crankshaft is controlled at the rear main bearing and the end clearance should be .002" to .004".

The proper end clearance for the connecting rods is .005" to .010".

GOVERNOR

The vacuum governor is regular equipment and the governor is set at 2600 R.P.M. maximum with a free running engine. This speed pulls down to about 2300 R.P.M. under load.

PARTS LIST No. 772

THE AUTOCAR COMPANY, ARDMORE, PENNSYLVANIA
SERVICE DEPARTMENT

BULLETIN REF. 585

ASSEMBLY HERCULES JXD (320) ENGINE - 4" x 4-1/4" - MODEL U-2044 CHASSIS

Part No.	Req.	Description	Hercules Motors Corp. Number
		CYLINDER AND CRANKCASE GROUP	
2UGA0809	1	Cylinder and Crankcase Assembly (Including Valve Seats, Main Bearings and Valve Guides)	42315-AS
2BL065	2	Valve Cover	40006-A
S-1871	4	Valve Cover Screw	4068-A
2UG0651	2	Main Bearing - Front	45693-B
2UG0653	8	Main Bearing - Intermediate	45696-B
2UG0655	2	Main Bearing - Center	45694-B
2UGQ657	2	Main Bearing - Rear	45695-B
2BL0521	As	Main Bearing Shim - Front & Inter. (.002)	40550-A
2BL0522	30	Main Bearing Shim - Front & Inter. (.003)	40552-A
2BL0523	As	Main Bearing Shim - Center & Rear (.002)	40549-A
2BL0552	12	Main Bearing Shim - Center & Rear (.003)	40551-A
S-922	10	Main Bearing Screw	40070-A
S-1349	10	Spring Washer	4731-A
S-3373	8	Main Bearing Screw	40071-A
S-1348	8	Spring Washer	4732-A
S-4212	9	Cylinder Welch Plug - 1-1/4"	1609-A
S-4208	1	Cylinder Oil Tube Plug	4643-A
S-629	1	Cylinder Oil Tube Pipe Plug - 1/4"	60-A
S-619	7	Cylinder Oil Tube Pipe Plug - 1/8"	59-A
S-2490	4	Cylinder Pipe Plug - 1/2"	4312-A
S-2500	1	Cylinder Drain Cock	8-A
2UG0210	1	Oil Level Gauge Assembly	45180-BS
2BL0946	1	Oil Filler Cap Assembly	40038-AS
2BLW0105	1	Oil Filler Body	40566-A
	1	Cylinder Oil Header Pipe Plug	22158-A
		VALVE GROUP	
2BL010	6	Valve-Exhaust	45814-A
2BL0710	6	Valve-Intake	40107-A
2BL052	12	Valve Spring	40008-A
2BL049	12	Valve Spring Cup	22013-A
2BL0329	12	Valve Stem Pin	40011-A
2BL021	12	Valve Guide	22011-A
2BL01001	6	Valve Seat Inserts	23123-A
2UG036	12	Valve Plunger	45812-A
2BL024	12	Valve Plunger Guide	22089-A
2TH0168	12	Valve Plunger Adjusting Screw	2185-A

Part No.	Req.	Description	Hercules Motors Corp. Number
		VALVE GROUP (Continued)	
2TH0748	12	Valve Plunger Adjusting Screw Lock Nut	2186-A
		CYLINDER HEAD GROUP	
2UG035	1	Cylinder Head	45921-D
S-3106	26	Cylinder Head Screw	4118-A
S-4172	1	Cylinder Head Pipe Plug - 1/2"	1075-A
		TIMING GEAR CASE COVER GROUP	
2UGA069	1	Timing Gear Case Cover Only	40339-D
S-1879	8	Timing Gear Case Cover Screw	4266-A
S-1347	8	Spring Washer	342-A
2BL0774	3	Timing Gear Case Cover Thrust Adjusting Screw Assembly	14594-AS
2BL01218	3	Timing Gear Case Cover Thrust Adjusting Screw Washer	14591-A
2TH0775	3	Timing Gear Case Cover Thrust Adjusting Screw Nut	2048-A
	1	Timing Gear Case Cover Oil Seal Assembly	11137-AS
	1	Timing Gear Case Cover Oil Seal	11088-A
		CONNECTING ROD GROUP	
2BL020	6	Connecting Rod Assembly	40390-CS
2BL0109	12	Connecting Rod Bearing Shell 45707-B or	45692-B
2BL0116	12	Connecting Rod Bearing Bolt	22059-A
2BL0118	12	Connecting Rod Bearing Bolt Nut	21056-A
2BL01189	18	Connecting Rod Bearing Shim (.003)	40556-A
2BL0553	As	Connecting Rod Bearing Shim (.002)	40553-A
2BL0162	6	Piston Pin Clamp Screw	22111-A
S-3185	6	Piston Pin Clamp Screw Lock Washer	14761-A
		PISTON GROUP	
2BN027	6	Piston-Std. (4" Dia. Aluminum)	40120-C
2BN0891	6	Piston-.010-Oversize (4" Dia. Aluminum)	40120-C
2BN0873	6	Piston-.020-Oversize (4" Dia. Aluminum)	40120-C
2BN0864	6	Piston-.030-Oversize (4" Dia. Aluminum)	40120-C
2BN039	6	Piston Pin	22238-B
	6	Piston Ring-Compression-Std. (4") (Top Groove)	42297-A
	6	Piston Ring-Compression-.010 Oversize (4") (Top Groove)	42297-A
	6	Piston Ring-Compression-.020 Oversize (4") (Top Groove)	42297-A
	6	Piston Ring-Compression-.030 Oversize (4") (Top Groove)	42297-A

Parts List No. 772
Bulletin Ref. #585

Part No.	Req.	Description	Hercules Motors Corp. Number
		PISTON GROUP (Continued)	
2BN025	12	Piston Ring-Compression-Std. (4") (2nd & 3rd Groove)	3813-A
2BN0888	12	Piston Ring-Compression-.010 Oversize (4") (2nd & 3rd Groove)	3813-A
2BN0875	12	Piston Ring-Compression-.020 Oversize (4") (2nd & 3rd Groove)	3813-A
2BN0353	12	Piston Ring-Compression-.030 Oversize (4") (2nd & 3rd Groove)	3813-A
2BN026	6	Piston Ring-Oil-Std. (4")	3913-A
2BN0889	6	Piston Ring-Oil-.010 Oversize (4")	3913-A
2BN0877	6	Piston Ring-Oil-.020 Oversize (4")	3913-A
2BN0878	6	Piston Ring-Oil-.030 Oversize (4")	3913-A
		CRANKSHAFT & FLYWHEEL GROUP	
2BL001	1	Crankshaft	45853-E
2BL0911	1	Crankshaft & Gear Assembly	
2BL023	1	Crankshaft Gear	22039-B
2BL01221	1	Crankshaft Dowel	4642-A
S-2319	1	Crankshaft Gear Key	4265-A
2BN006	1	Flywheel	23705-C
2BL0397	4	Flywheel Bolt	22104-A
S-811	4	Flywheel Bolt Nut	1656-A
2BL0214	1	Flywheel Gear (115 Teeth)	15471-C
2TH0837	2	Flywheel Dowel	1707-A
S-4207	2	Flywheel Welch Plug	665-A
2BL01225	1	Crankshaft Oil Thrower	40076-A
1BL037	1	Crankshaft Starting Ratchet	40806-A
		CAMSHAFT GROUP	
2UG054	1	Camshaft	45686-D
2BL076	1	Camshaft Gear	22049-B
S-2319	1	Camshaft Gear Key	4265-A
2UG0698	1	Camshaft Gear Nut	11023-A
2BL0776	1	Camshaft Plunger	40068-A
2TH0699	1	Camshaft Gear Thrust Washer	2045-A
2BL044	2	Camshaft Bearing-Front & Rear	40063-A
2BL043	2	Camshaft Bearing-Intermediate	40065-B
		CAMSHAFT IDLER GEAR GROUP	
2BL033	1	Camshaft Idler Gear	40135-B
2BL0652	1	Camshaft Idler Gear Thrust Washer	22107-A
2BL0183	1	Camshaft Idler Gear Shaft, Plunger and Plug Assembly	40136-A
2BL0213	1	Camshaft Idler Gear Shaft Bearing	40137-A
S-619	1	Camshaft Idler Gear Shaft Pipe Plug	59-A
2BL0776	1	Camshaft Idler Gear Plunger	40068-A

Parts List No. 772
Bulletin Ref. #585

Part No.	Req.	Description	Hercules Motors Corp. Number
		OIL PAN GROUP	
2UG0138	1	Oil Pan	42176-D
S-50	16	Oil Pan Screw	315-A
	4	Oil Pan Screw (Front End)	14770-A
S-1347	20	Spring Washer	342-A
S-4756	2	Oil Pan Pipe Plug	565-A
		OIL PUMP & OIL PRESSURE REGULATOR GROUP	
2UG0150	1	Oil Pump Assembly	45290-CS
2UG0129	1	Oil Pump Body	41124-C
2BL0131	1	Oil Pump Cover	40153-A
S-30	6	Oil Pump Cover Screw	14069-A
S-1345	6	Spring Washer	628-A
2BL0975	2	Oil Pump Snap Ring	4387-A
2BL0127	1	Oil Pump Shaft	22124-A
2BL0195	2	Oil Pump Shaft Gears	22154-A
S-2325	2	Oil Pump Shaft Gear Key	1179-A
2BL0126	1	Oil Pump Idler Shaft	22155-A
2BL0148	1	Oil Pump Drive Gear	45266-A
S-3192	1	Oil Pump Drive Gear Pin	4809-A
2BL0122	1	Oil Pump Drive Gear Washer	2047-A
2BL0598	1	Oil Pressure Regulator Valve	22129-A
2BL0477	1	Oil Pressure Regulator Spring	1347-A
2BL0956	1	Oil Pressure Regulator Screw	2058-A
S-3715	1	Oil Pressure Regulator Screw Nut	28-A
2BL01213	1	Oil Pressure Regulator Nut	1660-A
2BL01212	1	Oil Pressure Regulator Spring Button	1385-A
S-23	3	Oil Pump Attaching Screw	1864-A
S-1347	3	Spring Washer	342-A
2BL01215	1	Oil Pressure Crow Foot Wrench	2268-A
2BL01214	1	Oil Pressure "T" Wrench	X-5800
		MANIFOLD GROUP	
2UG0860	1	Intake & Exhaust Manifold	45806-E
2UG0159	10	Intake & Exhaust Manifold Stud	14895-A
S-1419	10	Intake & Exhaust Manifold Stud Nut	848-A
S-3326	10	Intake & Exhaust Manifold Stud Washer	1388-A
2BL01216	1	Intake & Exhaust Manifold Set Screw	4640-A
S-4720	1	Intake & Exhaust Manifold Pipe Plug	305-A
7BL004	1	Intake & Exhaust Manifold Flange	40314-A
		ENGINE SUPPORTS AND BRACKETS	
2UG0153	1	Rear Engine Support (Bellhousing #3 S.A.E.)	24476-D
2BL01268	1	Rear Engine Support Oil Seal	3543-B
S-2086	8	Rear Engine Support & Crankcase Screw	4123-A
S-20	3	Rear Engine Support Screw	2100-A
S-1349	10	Spring Washer	312-A
2BL2542	2	Rear Engine Support Cover Plate-Front	

Parts List No. 772
Bulletin Ref. #585

Part No.	Req.	Description	Hercules Motors Corp. Number
		ENGINE SUPPORTS AND BRACKETS (Continued)	
S-1461	6	Rear Engine Support Cover Plate Screw	
S-1347	6	Spring Washer	
2BL3543	1	Rear Engine Support Cover Plate-Rear-R. H.	
2BL3544	1	Rear Engine Support Cover Plate-Rear-L. H.	
2BL3132	1	Front Engine Support Bracket	40700-B
S-922	1	Front Engine Support Bracket Bolt	1701-A
S-1420	1	Front Engine Support Bracket Bolt Nut	3293-A
S-1349	1	Spring Washer	312-A
2BL4152A	1	Front Engine Support Cross Member	
2BL2277	1	Front Engine Support Rubber	
2BL2278	1	Front Engine Support Channel	
S-1877	2	Front Engine Support Channel Bolt	
S-1310	2	Front Engine Support Channel Bolt Nut	
12NJ2402	2	Rear Engine Support Rubber Biscuit	
12NJ2403	2	Rear Engine Support Rubber Bushing	
2BL2239A	2	Rear Engine Support Bolt	
S-808	2	Rear Engine Support Bolt Nut	
S-927	2	Rear Engine Support Bolt Washer	
		GASKET GROUP	
2BL072	2	Valve Cover Gasket	41005-B
2BL0178	1	Timing Gear Case Cover Gasket	22024-B
2BL0782	1	Rear Engine Support Gasket	22065-B
5BL004	1	Water Pump Gasket	22164-A
2BL0467	1	Water Pump Attaching Gasket	22149-A
2UG0161	2	Oil Pan Gasket (Half)	42183-B
2BL0578	1	Oil Pump Gasket	22119-A
2UG0633	1	Intake & Exhaust Manifold Gasket	45805-B
7BL022	1	Exhaust Manifold Flange Gasket	40028-A
2BN007	1	Cylinder Head Gasket	40237-C
6T2502A	1	Fuel Pump to Crankcase Gasket	
6BL2375	2	Governor Gasket	
2UBL2575	1	Oil Filter Attaching Gasket	
		AIR COMPRESSOR MOUNTING GROUP	
	1	Air Compressor Bracket	40565-C
	3	Air Compressor Bracket Screw	4354-A
	3	Spring Washer	312-A
	1	Air Compressor Oil Inlet Tube Assembly	40567-BS
	2	Air Compressor Oil Inlet Tube Nut	1532-A
	1	Air Compressor Oil Inlet Tube Union	1533-A
	1	Air Compressor Oil Inlet Tube Elbow	1659-A
	1	Air Compressor Oil Drain Tube Assembly	40568-BS
	2	Air Compressor Oil Drain Tube Nut	2397-A
	1	Air Compressor Oil Drain Tube Union	2396-A
	1	Air Compressor Oil Drain Tube Elbow	4086-A

Parts List No. 772
Bulletin Ref. #585

Part No.	Req.	Description	Hercules Motors Corp. Number
		AIR COMPRESSOR MOUNTING GROUP (Continued)	
	1	Air Compressor Driven Pulley	40511-B
	1	Air Compressor Water Inlet Tube Assembly	45366-BS
	1	Air Compressor Water Outlet Tube Assembly	45367-BS
	1	Air Compressor Water Tube Union	246-A
		MISCELLANEOUS GROUP	
	1	Ignition Tube Assembly	40109-B

SERVICE BULLETIN No. 689

THE AUTOCAR COMPANY, ARDMORE, PENNSYLVANIA

SERVICE DEPARTMENT

SUBJECT: GASKET KIT FOR HERCULES JXB AND JXC ENGINES - 2BL01200
GASKET KIT FOR HERCULES JXD ENGINE - 2BN01200

Part No.	Req.	Description
2BN007	1	Cylinder Head Gasket (Used only on JXD Engine)
2BL007	1	Cylinder Head Gasket (Used only on JXB and JXC Engines)
2BL072	2	Valve Cover Gasket
2BL0161	2	Oil Pan Gasket (Half)
2BL0178	1	Timing Gear Case Cover Gasket
2BL0467	1	Water Pump Attaching Gasket
2BL0575	1	Oil Filter Attaching Gasket (Use with Refillable Type Filter Element)
2UBA0575	1	Oil Filter Attaching Gasket (Use with Cartridge Type Filter Element)
2UBL2575	1	Oil Filter Attaching Gasket (Use with Cartridge Type Filter Element)
2BL0578	1	Oil Pump Gasket
2BL0621	1	Timing Gear Case Cover Cork
2BL0633	1	Intake & Exhaust Manifold Gasket
2BL0634	1	Oil Pan Cork Seal
2BL0782	1	Rear Engine Support Gasket
2SA0828	1	Oil Filter Shell Gasket (Use with Refillable Type Filter Element)
2UA0828	1	Oil Filter Element Gasket (Use with Cartridge Type Filter Element)
5BL004	1	Water Pump Gasket
6BL2375	2	Governor Gasket
6T2502A	1	Fuel Pump to Crankcase Gasket
7BL022	1	Exhaust Manifold Flange Gasket
2BL01200	1	Gasket Kit - Complete for Hercules JXB and JXC Engines
2BN01200	1	Gasket Kit - Complete for Hercules JXD Engine

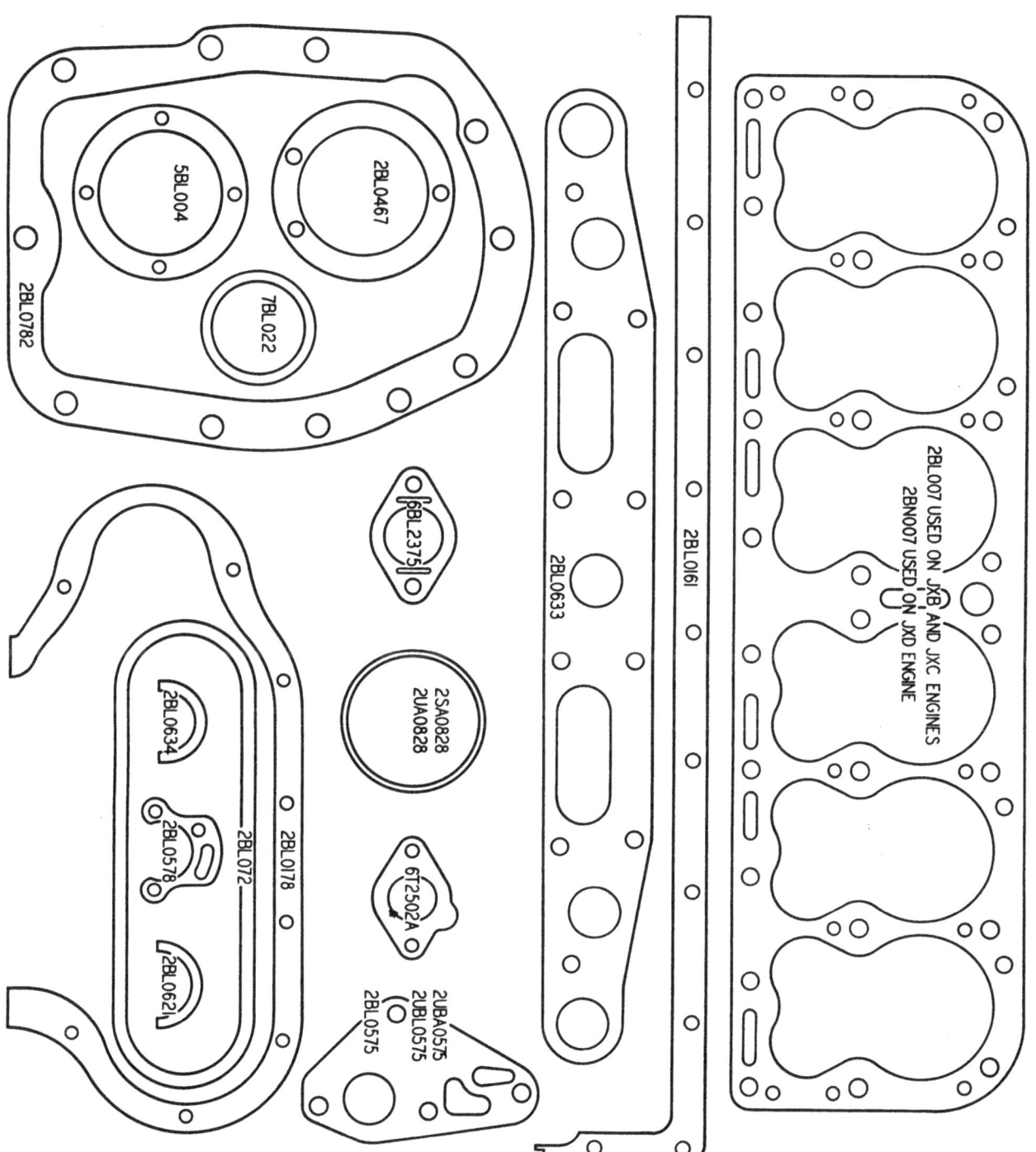

SERVICE BULLETIN   No. 599

THE AUTOCAR COMPANY, ARDMORE, PENNSYLVANIA
SERVICE DEPARTMENT

SUBJECT: DUO-FLO OIL FILTER WITH CARTRIDGE TYPE FILTER ELEMENT- ON JXB AND JXC HERCULES ENGINES

This unit is equipped with a cartridge type filter which is very efficient in removing sludge as well as carbon and other foreign materials from the oil. The filtering element also has the property of neutralizing any acid in the oil.

The filter cartridge, part number 2UA0807, should be replaced at periods of 5,000 to 10,000 miles, depending upon the nature of service. The cartridge can be replaced at very low cost. DO NOT ATTEMPT TO CLEAN AND REUSE AN OLD CARTRIDGE. It is loaded with dirt and other foreign substances some of which cannot be dissolved in gasoline. The cartridge should be changed when oil begins to darken on bayonet gauge.

To remove the cartridge, unscrew the handle on the top of the filter unit, release the lock bar from the upright tie rods and lift out.

Refill crankcase to proper level plus one quart for the filter. Run motor and check for leaks.

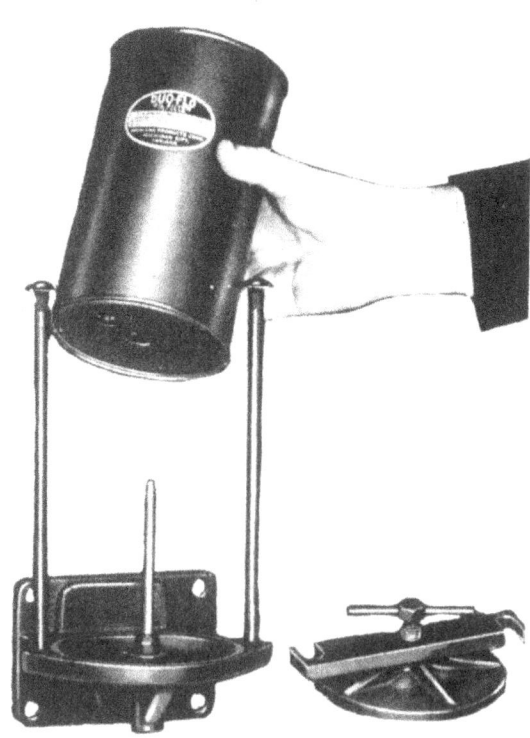

PARTS LIST

No. 689

THE AUTOCAR COMPANY, ARDMORE, PENNSYLVANIA
SERVICE DEPARTMENT

BULLETIN REF. 599

ASSEMBLY

DUO-FLO OILFILTER WITH CARTRIDGE TYPE FILTER ELEMENT-
CAB-OVER-ENGINE MODELS WITH JX HERCULES ENGINES

Part No.	Req.	Description	Michiana No.
2UBA0808	1	Oil Filter Assembly	2650-20
		BASE CASTING GROUP	
2UB5727B	1	Oil Filter Base Casting	2651-20
2UBA0575	1	Oil Filter Attaching Gasket	2650-20-G
2UA01204	1	Oil Filter Restriction Plug	1230
S-1672	1	Oil Filter By-Pass Valve Ball	520
2UBA0726	1	Oil Filter By-Pass Valve Spring	12018
2UBA0725	1	Oil Filter By-Pass Valve Plug	12557
2T0831	1	Oil Filter By-Pass Valve Plug Gasket	555
S-1428	2	Oil Filter Base Screw-Long	
S-19	2	Oil Filter Base Screw-Short	
S-1349	4	Spring Washer	
		FILTER ELEMENT AND COVER GROUP	
2UA0807	1	Oil Filter Element (Cartridge Type)	12550
2UA0318	1	Oil Filter Cover	12256
2UA0828	1	Oil Filter Element Gasket	12254
2UA0319	1	Oil Filter Cover Wing Screw	12730
2UA01203	1	Oil Filter Cover Wing Screw Rod	12732
2UA0608	1	Oil Filter Cover Cross Bar	12733
2UBA01202	2	Oil Filter Element Tie Rods	12731-B

SERVICE BULLETIN No. 700

THE AUTOCAR COMPANY, ARDMORE, PENNSYLVANIA
SERVICE DEPARTMENT

SUBJECT GOVERNOR - HANDY VELOCITY TYPE, MODEL 703-722-138

CONSTRUCTION

The Model 703-722-138 Governor (Autocar 2UK0650) is a velocity-vacuum type which is set for an engine speed of 2400 R.P.M. at full load, corresponding to a maximum of 2700 R.P.M. with no load.

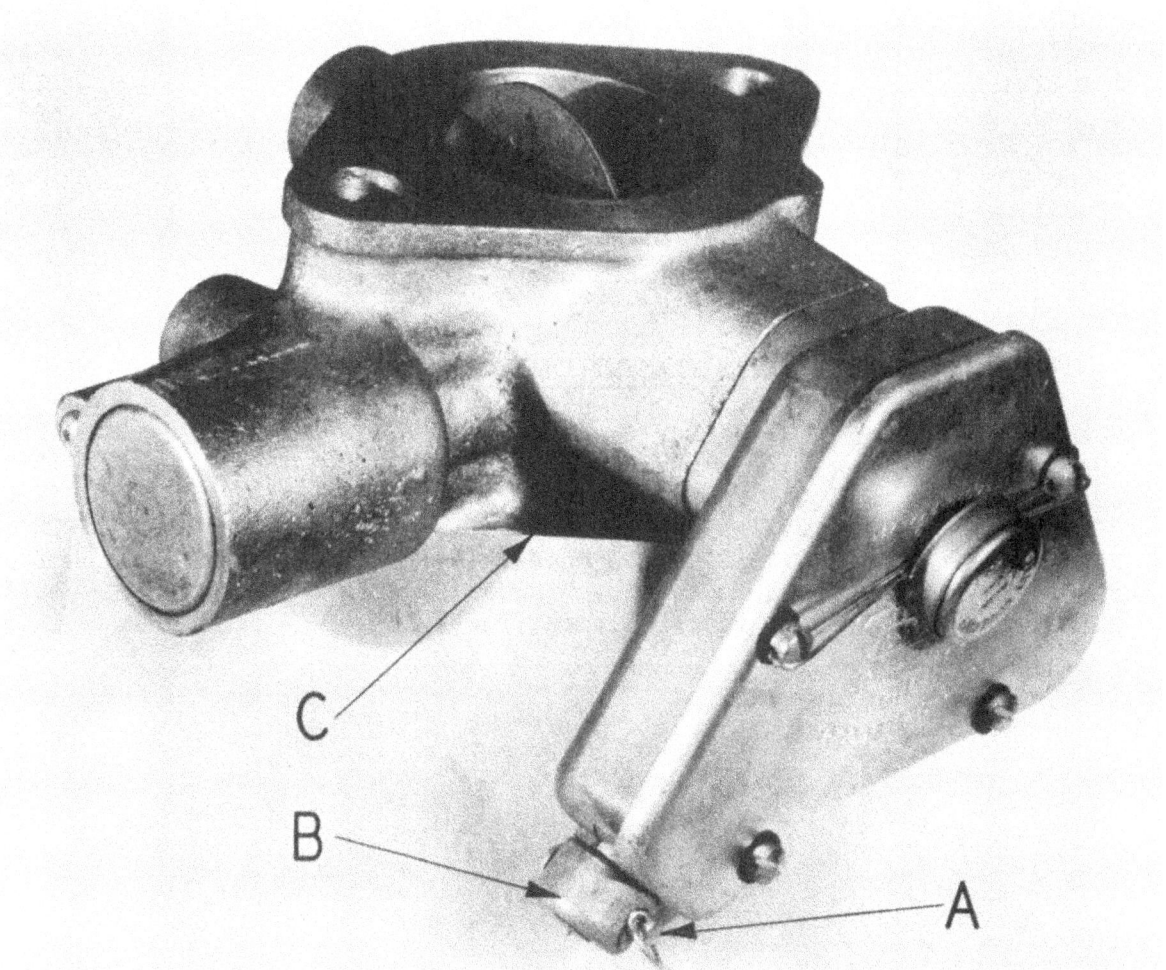

INSTALLATION

This governor is installed between the carburetor and intake manifold so that adjusting cap B is accessible. Flange C marked "carburetor side" must be applied against the carburetor flange, even though the name plate may be upside down in this position.

ADJUSTMENT

Remove seal wire and pin at A. With wide open throttle, turn adjusting cap B to left for lower speed or right for higher speed, until desired engine speed is obtained. Replace pin in adjusting cap and seal. This Governor requires no lubrication.

PARTS LIST

No. 866

THE AUTOCAR COMPANY, ARDMORE, PENNSYLVANIA
SERVICE DEPARTMENT

BULLETIN REF. 700

ASSEMBLY 2UK0650 - HANDY VELOCITY TYPE GOVERNOR, MODEL 703-722-138

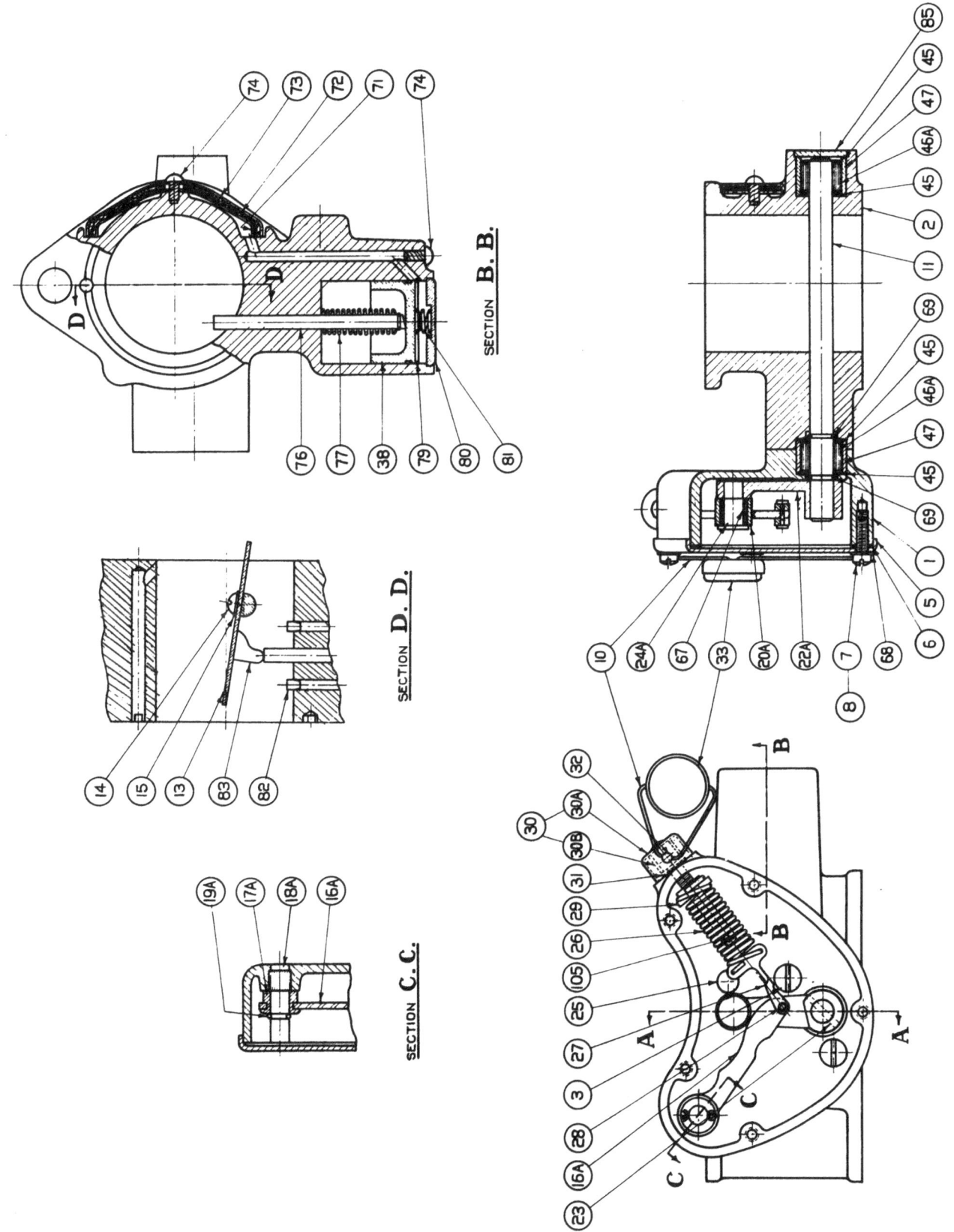

PARTS LIST No. 866

THE AUTOCAR COMPANY, ARDMORE, PENNSYLVANIA
SERVICE DEPARTMENT   BULLETIN REF. 700

ASSEMBLY   2UK0650 - HANDY VELOCITY TYPE GOVERNOR 703-722-138

IMPORTANT: ALWAYS SPECIFY ITEM NUMBER AND GOVERNOR MODEL OR SERIAL NUMBER WHEN ORDERING PARTS

Item Number	Req.	Description
1	1	Housing (Not Sold Separately-Replace Complete Governor)
2	1	Flange (Not Sold Separately-Replace Complete Governor)
3	2	Housing to Flange Screw
5	1	Cover
6	1	Cover Gasket
7	3	Cover Screw (Drilled)
8	2	Cover Screw (Plain)
10	1	Cover Screw Seal
11	1	Throttle Shaft (Not Sold Separately-See Item #63)
13	1	Throttle Plate (Not Sold Separately-See Item #64)
14	2	Throttle Plate of Cam Fulcrum Screw
15	2	Throttle Plate Screw Lock Washer
16-A	1	Cam (Not Sold Separately - See Item #66)
17-A	1	Cam Bushing
18-A	1	Cam Fulcrum
19-A	1	Cam Fulcrum Screw Washer
20-A	1	Cam Roller
22-A	1	Cam Roller Lever (Not Sold Separately-See Item #63)
23	1	Cam Roller Lever Pin
24-A	1	Cam Roller Lever Shaft
25	1	Cam Roller Stop Pin
26	1	Spring
27	1	Spring Link (Use New Pin #28 with this Part)
28	1	Spring Link Pin
29	1	Spring Adjusting Block
30	1	Adjusting Screw Assembly
31	1	Adjusting Screw Washer
32	1	Adjusting Screw Seal Pin
33	1	Adjusting Screw Seal
38	1	Stabilizer Piston
45	4	Throttle Shaft Bearing Shim Washer
46-A	22	Throttle Shaft Roller Bearing
47	2	Throttle Shaft Roller Bearing Sleeve
63	1	Throttle Shaft Assembly (Consists of Items 11, 20-A, 22-A, 23, 24-A and 67)
64	1	Throttle Plate Assembly (Consists of Items #13, 83)
66	1	Cam Assembly (Consists of Items 16-A, 17-A, 27, 28)
67	1	Cam Roller Bearing
68	2	Cover Screw Lock Washer
69	2	Throttle Shaft Bearing Retaining Ring
71	1	Air Cleaner Plate (Inner)
72	1	Air Cleaner Felt

Parts List No. 866
Bulletin Ref. #700

IMPORTANT: <u>ALWAYS SPECIFY ITEM NUMBER AND GOVERNOR MODEL OR SERIAL NUMBER WHEN ORDERING PARTS</u>

Item Number	Req.	Description
73	1	Air Cleaner Cover (Outer)
74	1	Air Cleaner Cover Drive Screw
76	1	Piston Rod
77	1	Piston Rod Spring
79	1	Piston Stop Ring
80	1	Cylinder Plug
81	1	Auxiliary Spring
83	1	Throttle Plate Arm
84	1	Throttle Shaft Roller Bearing Sleeve
85	1	Throttle Shaft Hubbard Plug
105	1	Adjusting Screw Wire

SERVICE BULLETIN No. 633

THE AUTOCAR COMPANY, ARDMORE, PENNSYLVANIA
SERVICE DEPARTMENT

SUBJECT: CLUTCH ASSEMBLY - MODEL 12 CB-C

Beginning with engines 10-827 and 20-4748, the Model 12CB-C clutch assemblies for the RM and RL Models will be built with an improved release lever yoke, spring and adjusting nut construction.

The new yokes have been enlarged on the shank diameter from 3/8" to 7/16" but will interchange with the earlier construction if new nuts and springs are used, and also providing that the diameter of the holes in the cover plate are increased from 1/2" to 9/16".

Replacements should be made in sets including all six yoke assemblies rather than any one or two sets of parts.

CONSTRUCTION   This clutch is of the single disc type, using one driven disc assembly 1 attached to a splined hub which is mounted on the clutch shaft 35. The clutch cover 6 contains the pressure plate 9, springs 17 and the release fingers 13.

The clutch is released by action of the foot pedal which moves the trunnion block 19 against the inner ends of the six release fingers 13. The release fingers mounted to the pressure plate 9 and cover plate 6 on needle bearings, pull back on the clutch release pins 10 which in turn pull back on the pressure plate 9 compressing the pressure spring 17.

LUBRICATION - Trunnion Block Bearing. The trunnion bearing 23 is lubricated through a tube connected to the wick feed oiler mounted on the water jacket plate on the left hand side of the engine. Refill reservoir with light oil weekly.

Clutch Shaft Pilot Bearing. The clutch shaft pilot bearing 24 is packed with grease when the clutch is assembled and does not require further attention except when rebuilding the unit, at which time it should be repacked with grease.

ADJUSTMENTS - Normal Wear. The pressure springs will automatically compensate for all wear on the friction facings without any internal adjustments to the clutch. It is essential, however, to adjust the pedal travel as normal wear occurs and the clearance between the pedal and floorboard is reduced. If this is neglected, clutch slippage will result.

Adjust the wing nut on the link between the clutch control shaft lever and the clutch pedal lever so that with the clutch engaged there is approximately 1-1/2" free travel of the pedal pad before the trunnion block contacts the release fingers.

Page 18

Caution - Release finger adjusting nuts 15 are adjusted at the Factory and must not be disturbed during life of clutch except after relining of friction discs.

SERVICE HINTS

Disassembling from Flywheel - Remove transmission assembly exposing clutch and flywheel.

Remove the 12 cap screws 25 which attach the clutch cover 6 to the flywheel, unscrew each a few turns at a time so the release of the spring load is nearly equal all around.

Servicing Cover Plate Assembly - It is important that the relation of the cover plate 6 to the pressure plate 9 be marked so that they can be reassembled accordingly or the running balance of the clutch may be affected. Particular attention should be paid to the position of all original parts during the tear down operation so that replacement may be made correctly.

If an arbor press is available, place the cover assembly 6 on the bed of the press with a 2 x 4 block of wood under the pressure plate 9 so arranged that the cover 6 is free to move down. Place a block or bar across top of cover between release lever adjusting nuts 15.

Compress the cover 6 with the spindle of the press and holding it under compression, remove the six release lever adjusting nuts 15 and then slowly release the pressure until cover plate stamping can be removed, making all parts available for inspection.

If arbor press is not available three "C" clamps will answer the purpose.

Inspect all parts carefully. If the pressure plate 9 is discolored from excessive heat, the pressure springs 17 should be checked for load if less than 140# @ 1-9/16" they should be replaced.

If pressure plate 9 is grooved, warped or heat checked, it should also be replaced.

When new pressure plate is installed or if release levers 13 or release lever yokes 14 are worn, disassembly of these parts will be necessary. Before reassembling, wash parts in gasoline and wipe dry. A dummy pin is necessary to assemble the needle rollers 12. It can be made from a release lever pin 11 by sawing off the head so it is as long as the lever is wide and chamfering the ends, the needle rollers can be assembled around the dummy pin. Place the yoke 14 in position on the lever and force the dummy pin out with the lever pin - short 11 and lock with cotter pin. Repeat this operation to install needle rollers 12 to lever and place in position on pressure plate 9, using lever pin - long 10 locking same with cotter pin.

Reassemble the insulator buttons 18, pressure springs 17 and set release lever yokes 14 vertical to the pressure plate, placing tension springs 16 in position on yokes 14. Mount the cover over the pressure springs making sure that all parts are properly located and that the marks made on pressure plate and cover plate before disassembling are lined up. Place an arbor press with blocks under pressure plate and bar across top as in dismantling and slowly compress, guiding pressure plate lugs through windows in cover plate and release lever yokes through holes in top of cover.

Assemble release lever adjusting nuts 15 and tighten down about three threads below the head of the bolt.

The clutch should now be released several times to make sure that all moving parts are working freely. This can be done by applying the spindle of the press to the inner ends of the release levers.

Bulletin #633

Servicing Driven Member Assembly - The driven member has six cushion plates placed between the disc 1 and rear facing for the purpose of giving smooth engagement and prolonging facing wear and under no circumstances should they be left out.

To disassemble - center punch iron rivets 3 on flywheel side - (or short end of hub side). Spot drill iron rivets with 3/16" and punch out, being careful that the driven member is well supported so as not to distort the disc. Remove facing 4 from cushion plates and disc 1 by punching out the rivets 5. To reassemble - rivet one facing to disc on flywheel side (or short end of hub side) with brass tubular rivets 5. Rivet cushion plates to other facing with rivets 5. Then rivet tightly facing and cushion plate unit to disc with iron rivets 3.

Caution - Brass rivets must not extend through disc or cushion plates more than 1/32". After refacing the driven member, it should be checked if possible for runout. This should not exceed .035 measured near the outside edge of the front facing.

Assembling Clutch in Flywheel - The splines of the shaft should be coated with a little heavy oil. Do not use too much as it will be thrown off on the facings. Inspect the pilot bearing and release bearing to make sure they are in good condition and well lubricated with a good grade of high melting point grease of medium viscosity. The driven member should first be placed in the flywheel, then the cover plate assembly. Next a line up shaft should be pushed through the clutch hub into the pilot bearing. If a dummy shaft is not available, an extra transmission shaft can be used. Insert the 12 pilot cap screws 25 and tighten uniformly, giving each a few turns at a time until all are seated. Remove dummy shaft and adjust levers.

Lever Adjustment - The 6 release levers 13 must be adjusted exactly to 31/32" from a straight edge laid across the top surface of the cover plate to tip of release levers, and should be in plane within 1/32". Lock the adjusting nuts 15 with a staking chisel or screw driver, peening portions of the nut into the slot of the bolt.

Before assembling the transmission to the engine, see that the release sleeve 19 slides freely on the extension of the transmission driveshaft cap 36, and the oil line for the sleeve is connected and not damaged.

PARTS LIST

No. 645

THE AUTOCAR COMPANY, ARDMORE, PENNSYLVANIA
SERVICE DEPARTMENT

BULLETIN REF. 633

ASSEMBLY   12CB-C LONG CLUTCH ASSEMBLY - 3RL430A

Sketch Ref. Number	Part No.	Req.	Description	Long Mfg. Co's. No.
	3RL430A	1	Clutch Assembly (Does not include Housing and Control)	5854
			DRIVEN MEMBER GROUP	
1	3RL0770	1	Driven Member Assembly (Including Hub and Plate Assembly)	CM-4232
2	3RL01470	1	Cushion Plates & Facing Unit Assembly	CS-3599
3	3RL0785	12	Cushion Plates & Facing Unit Rivet	C-3006
4	3RL045	2	Friction Facing	C-3593
5	3RL0114	48	Friction Facing Rivet	C-3429
			COVER PLATE GROUP	
6	3RL0800A	1	Cover & Pressure Plate Assembly (Including Cover Plate, Springs & Pressure Plate)	CM-4348
6	3RL0889A	1	Cover Plate Unit (Including Riveted Spring Cups)	CS-4349
6	3RL048A	1	Cover Plate-Bare	C-4343
7	3RL0642	12	Spring Cup	C-3437
8	3RL0644	12	Spring Cup Rivet	C-3436
9	3RL0790A	1	Pressure Plate & Lever Unit	CS-4347
9	3RL043	1	Pressure Plate-Bare	C-3594
10	3RL0782	6	Release Lever Pin	C-3141
11	3RL0783	6	Release Lever Adjusting Yoke Pin	C-3142
12	3RL0781	156	Needle Roller	C-3077
13	3RL046	6	Release Lever	C-3390
14	3RL0784A	6	Release Lever Adjusting Yoke	C-4344
15	3RL0105A	6	Release Lever Adjusting Nut	C-4345
16	3RL0132A	6	Release Lever Tension Spring	C-4346
17	3RL0128	12	Pressure Spring	C-3431
18	3T0124B	12	Pressure Spring Insulator Button	C-2085
19	3A0780	1	Release Sleeve Assembly (Including Wick & Spring Post)	CM-1012
19	3A0125	1	Release Sleeve-Bare	C-1888
20	3A0145	1	Oil Wick	C-860
21	3A0136	1	Return Spring Post	C-959
22	10W165	1	Return Spring	
	10R1148	1	Return Spring Link	
23	3BM2169A	1	Clutch Trunnion Bearing	A-959-1
24	3D0243	1	Clutch Pilot Bearing	6304-Z
25	S-50	12	Cover Plate Screw	
	S-1347	12	Spring Washer	

Parts List No. 645
Bulletin Ref. #633

Sketch Ref. Number	Part No.	Req.	Description	Long Mfg. Co's. No.
			CLUTCH HOUSING & CROSS SHAFT GROUP	
26	3D0580	1	Clutch Housing & Bushings Assembly (S.A.E. #2)	
26	3D5343	1	Clutch Housing (S.A.E. #2)	
27	S-3127	10	Clutch Housing Screw	
	S-1349	10	Spring Washer	
	3UN2119A	2	Clutch Housing Bushing	
28	3T352A	1	Clutch Housing Inspection Plate	
29	S-12	2	Clutch Housing Inspection Plate Screw	
	3Y2113	1	Flywheel Sight Hole Cover Plate	
30	10DF361A	1	Clutch Control Cross Shaft	
31	10T258	2	Clutch Trunnion Lever	
32	S-2311	2	Clutch Trunnion Lever Key	
33	S-64	2	Clutch Trunnion Lever Screw	
	S-1347	2	Spring Washer	
	10L01330	1	Clutch Control Shaft & Control Lever Assembly	
	10NL309	1	Clutch Control Lever	
	S-2316	1	Clutch Control Lever Key	
	S-923	1	Clutch Control Lever Screw	
	S-1348	1	Spring Washer	
	S-1347	2	Spring Washer	
			CLUTCH PEDAL BRACKET GROUP	
	10NL0828	1	Pedal Bracket & Bushing Assembly	
	10NL4138	1	Pedal Bracket	
	10NL215	2	Pedal Bracket Bushing	
	S-1872	4	Pedal Bracket Bolt	
	S-77	4	Pedal Bracket Bolt Nut	
	S-1349	8	Spring Washer	
	10NL255A	1	Pedal Lever Shaft	
	10Y1135	1	Pedal Lever Shaft Collar	
	S-2208	1	Cotter Pin	
	10NL453A	1	Clutch Pedal	
	S-2316	1	Clutch Pedal Key	
	S-40	1	Clutch Pedal Bolt	
	S-75	1	Clutch Pedal Bolt Nut	
	S-1347	2	Spring Washer	
	10UN2292	1	Clutch Pedal Pad	
	S-3086	2	Clutch Pedal Pad Bolt	
	S-1345	2	Spring Washer	
	10NL0833	1	Clutch Pedal Lever & Bushing Assembly	
	10NL208	1	Clutch Pedal Lever	
	10UA115	1	Clutch Pedal Lever Bushing	
	S-2316	1	Clutch Pedal Lever Key	
	S-923	1	Clutch Pedal Lever Screw	

Parts List No. 645
Bulletin Ref. #633

Sketch Ref. Number	Part No.	Req.	Description	Long Mfg. Co's. No.
			CLUTCH PEDAL BRACKET GROUP (Continued)	
	S-1348	1	Spring Washer	
	S-1880	1	Clutch Pedal Lever Stop Screw	
	S-77	1	Clutch Pedal Lever Stop Screw Nut	
	10DFL0470	1	Clutch Control Rod Assembly	
	10DFL248	1	Clutch Control Rod (6-1/2" Long)	
	S-1418	1	Clutch Control Rod Nut	
	10A256	1	Clutch Control Rod Clevis	
	S-2259	1	Clutch Control Rod Clevis Pin	
	S-2198	1	Cotter Pin	
	10A2383	1	Clutch Control Rod Adjusting Clevis	
	S-2257	1	Clutch Control Rod Adjusting Clevis Pin	
	S-2198	1	Cotter Pin	
	10T179A	1	Clutch Pedal Return Spring Link- Upper	
	10S2147	1	Clutch Pedal Return Spring	
	10CH279A	1	Clutch Pedal Return Spring Link- Lower	
			CLUTCH TRUNNION OILER GROUP	
	3CA91020	1	Clutch Trunnion Oil Line Assembly (1-1/4" I.D.)	
	3UN2625	1	Clutch Trunnion Oil Line Clamp Nut	
	S-4501	1	Clutch Trunnion Oil Line Ell	
	S-5023	1	Clutch Trunnion Oil Line Single Union	
	S-4302	1	Clutch Trunnion Oil Line Union Nut	
	3B3298	1	Magazine Oiler	
	3M2301A	1	Magazine Oiler Wick	
	2TE21008C	2	Magazine Oiler Stud	
	S-829	2	Magazine Oiler Stud Nut	
	S-1347	4	Spring Washer	
	S-4131	1	Magazine Oiler Filler Plug - 3/4"	
	S-619	3	Magazine Oiler Plug - 1/8"	

SERVICE BULLETIN   No. 682

THE AUTOCAR COMPANY, ARDMORE, PENNSYLVANIA
SERVICE DEPARTMENT

SUBJECT UK 4-SPEED TRANSMISSION - MODEL 231F-8

CONSTRUCTION

The Model UK transmission is a unit with 4 speeds forward and 1 reverse, with direct drive in fourth. This unit is equipped with helical gears running in constant mesh to provide quiet operation in third speed and direct drive.

The transmission main unit is illustrated in Fig. 1, the gearshift forks and selector finger mechanism in Fig. 4 and the gearshift hand lever control and connecting rod in Fig. 2.

The main shaft 1 is supported at the rear by an annular ball bearing 12 which takes both the thrust and radial load, and at the front by a roller bearing 11 carried in the end of the transmission drive gear 29. The drive gear is supported in the transmission case by an annular ball bearing 30 and is piloted at the front in the crankshaft by an annular ball bearing. Needle bearings are used in mounting the third speed constant mesh gear 3 on the main shaft.

The countershaft 15 is supported at the front end by a straight type roller

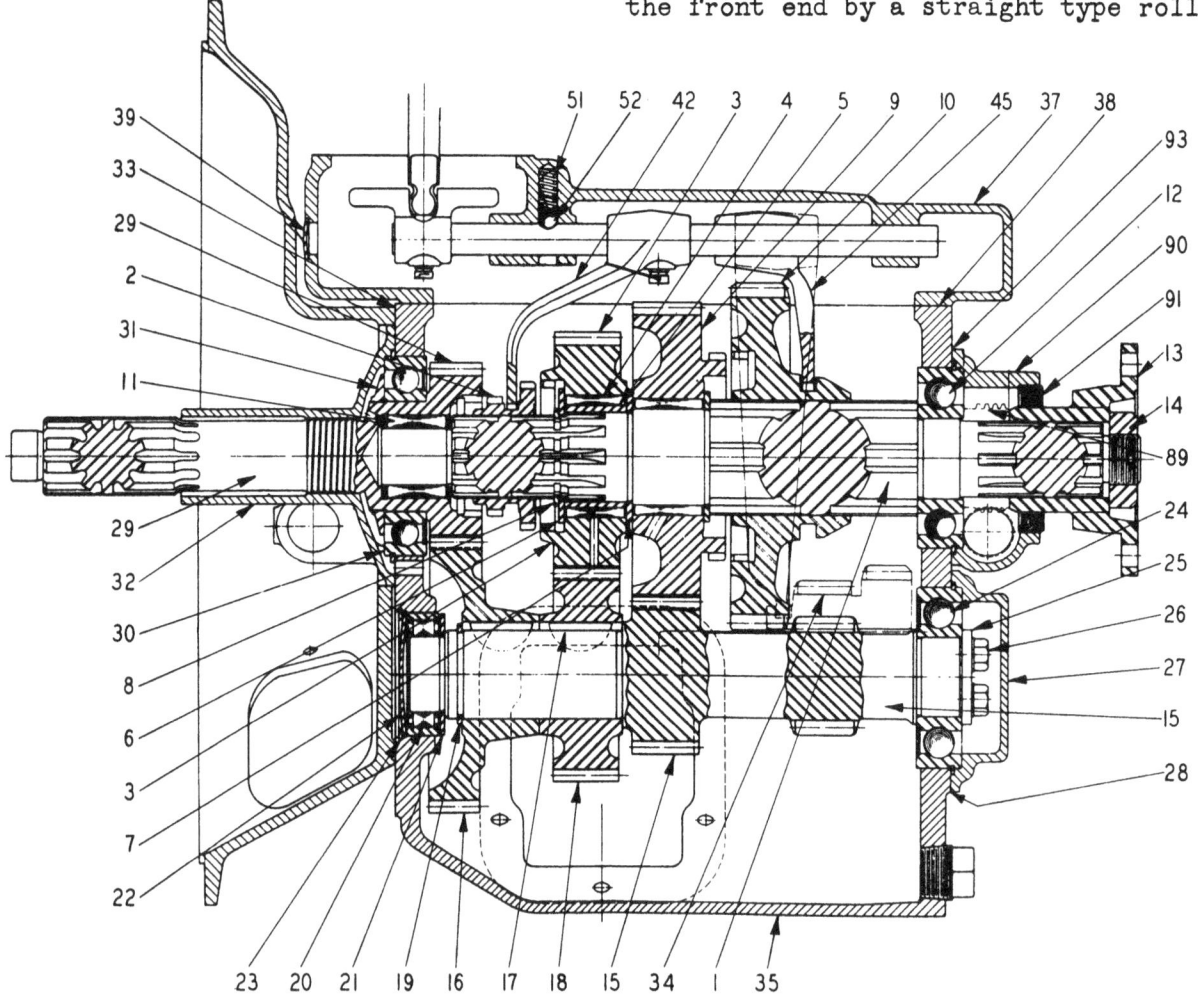

Fig. 1

Bulletin 682

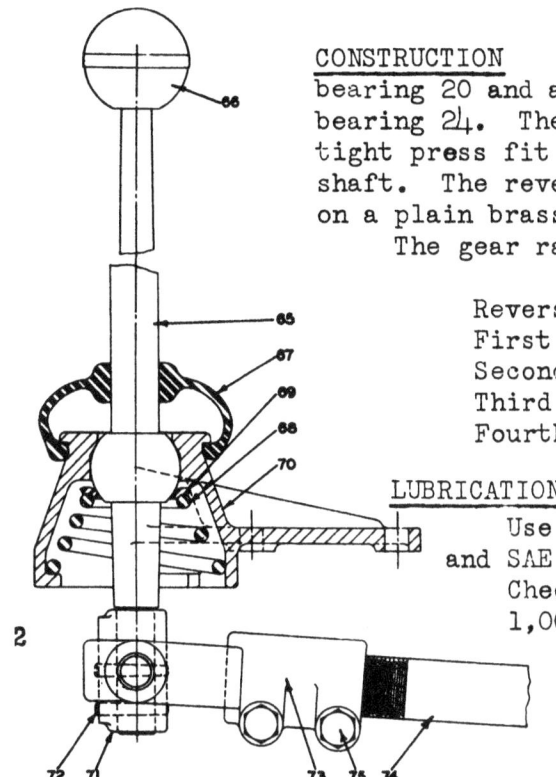

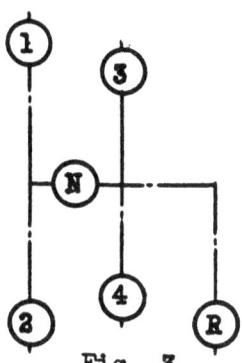

CONSTRUCTION

bearing 20 and at the rear by an annular ball bearing 24. The countershaft gears are a tight press fit and keyed on to the countershaft. The reverse idler gear 34 is carried on a plain brass bushing.

The gear ratios are as follows:

Reverse	7.41 to 1
First	6.35 to 1
Second	3.90 to 1
Third	1.97 to 1
Fourth (Direct)	1.00 to 1

Fig. 3

LUBRICATION

Use SAE No. 140 gear oil for summer operation and SAE No. 90 for winter operation.

Check level of lubricant weekly or every 1,000 miles and add oil as required.

Drain case, wash out with kerosene and refill every three months or every 10,000 miles. Capacity 5-1/2 quarts.

POWER TAKE-OFF

A standard SAE large opening is provided on both sides of the transmission for power take-off equipment. The drive is taken through an adapter in either case which is fitted with a driving gear having 15-teeth of 6-8 pitch revolving at .656 times engine speed in the same direction as crankshaft rotation.

OPERATION

The gearshift is selective. Gearshift lever positions are shown diagrammatically in Fig. 3.

A lockout device 48 is provided to prevent unintentional engagement of the reverse speed gear.

DISASSEMBLY

To completely disassemble transmission, it is necessary to remove unit from car and then remove top cover, speedometer drive gear case, driveshaft and rear countershaft bearing caps, then proceed as follows:

(a) Remove the mainshaft rear bearing from the case and shaft, and then by slipping the mainshaft to the rear, mainshaft assembly can be removed from the case.

(b) Remove the transmission drive gear and bearing from the front.

(c) Remove the thrust plate 22 from the front end of the countershaft.

(d) Remove the countershaft rear bearing.

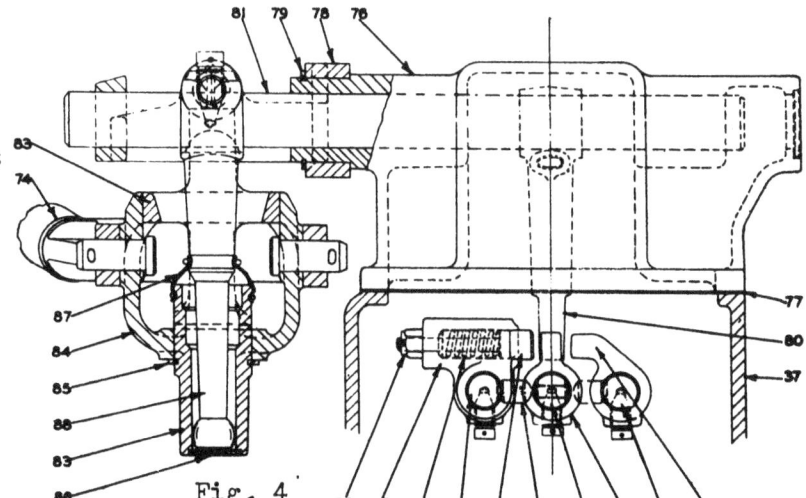

Fig. 4

(e) Slip the countershaft assembly to the rear and remove from the case.

(f) Remove reverse idler gear shaft lock from the rear end of case and remove shaft and gear.

The reverse procedure should be used when reassembling. If the main shaft third speed bearing retainer ring 8 is removed from the third speed gear, a new retainer ring must be used when reassembling. Never attempt to use the original retainer ring the second time since distortion of this ring during the removal operation can not be prevented. Failure to always use a new ring may result in breakage of the ring when in service and cause damage to the transmission gears.

PARTS LIST

No. 830

THE AUTOCAR COMPANY, ARDMORE, PENNSYLVANIA
SERVICE DEPARTMENT

BULLETIN REF. 682

ASSEMBLY   UK - FOUR SPEED TRANSMISSION - CAB-OVER-ENGINE MODELS

Sketch Ref. Number	Part No.	Req.	Description	Clark Equip. No. (Except as Noted)
	3UK6530	1	Transmission Assembly (Complete with Clutch Housing and Hand Control)	231-F-8
		1	Transmission Assembly (with Clutch Housing and Less Hand Control) (Part of 3UK6530)	
		1	Transmission Gearshift Control Assembly (Complete with Gearshift Lever and Bracket, Connecting Rod, Universal Lever and Top Control Cover) (Part of 3UK6530)	
			MAINSHAFT GROUP	
1	3UK031	1	Mainshaft - Bare	2681
2	3BL0951	1	Third & Fourth Speed Shift Hub	1064
3	3UK018	1	Mainshaft Third Speed Gear	3197
4	3BLO1016	36	Mainshaft Third Speed Gear Rollers	1089
5	3BLO1017	1	Mainshaft Third Speed Gear Roller Bushing	1073
6	3BLO1014	1	Mainshaft Third Speed Gear Roller Bearing Retainer Washer	1023
	3BLA01044	As	Mainshaft Third Speed Gear Shim .003	2008
	3BLA01045	As	Mainshaft Third Speed Gear Shim .005	2009
7	3BLO1013	1	Mainshaft Third Speed Gear Locating Washer	1020
8	3BL0957	1	Mainshaft Third Speed Gear Roller Bearing Retaining Ring	1051
	3BLO1015	1	Mainshaft Third Speed Gear Bushing Lock Pin	1052
9	3UK0102	1	Mainshaft Second Speed Gear	2685
	3BLA01016	34	Mainshaft Second Speed Gear Rollers	1003
	3BLA01013	1	Mainshaft Second Speed Gear Locating Washer	1019
10	3UK022	1	First & Reverse Slide Gear	2687
11	3BLA0147	1	Mainshaft Pilot Bearing	94622 (Hyatt)
12	3BLA0251	1	Mainshaft Rear Bearing	43308 (New Dept.)
13	3UL0201	1	Mainshaft Driving Flange	4855
14	3BLA0103	1	Mainshaft Driving Flange Nut	1211

Parts List No. 830
Bulletin Ref. #682

Sketch Ref. Number	Part No.	Req.	Description	Clark Equip. No. (Except as Noted)
			COUNTERSHAFT GROUP	
15	3UK029	1	Countershaft	2686
16	3BL021	1	Countershaft Drive Gear	1067
17	S-2303	2	Countershaft Gear Keys	4A-D
18	3UK026	1	Third Speed Countershaft Gear	3196
19	3BL01012	1	Countershaft Gear Retaining Ring	1910
20	3BL0142	1	Countershaft Front Bearing	1206TS (Hyatt)
21	3BL0916	1	Countershaft Front Bearing Locating Washer	1099
22	3BL01031	1	Countershaft Front Bearing Welch Plug	1560
23	3BL01003	1	Countershaft Front Bearing Welch Plug Retaining Ring	1901
24	3UK0194	1	Countershaft Rear Bearing	308MFG (Marlin Rockwell)
25	3BLA01038	1	Countershaft Rear Bearing Retaining Washer	1027
26	S-1783	2	Countershaft Rear Bearing Retaining Washer Screw	8A-501
	S-1347	2	Spring Washer	1A-600
27	3UK0977	1	Countershaft Rear Bearing Cap	2717
	S-23	4	Countershaft Rear Bearing Cap Screw	7A-608
	S-1347	4	Spring Washer	1A-600
28	3UK0156	1	Countershaft Rear Bearing Cap Gasket	2747
			DRIVESHAFT GROUP	
29	3UBT041	1	Transmission Driveshaft and Gear	1164
30	3BLA079	1	Transmission Driveshaft Bearing	47511 (New Dept.)
31	3BLA0994	1	Transmission Driveshaft Bearing Retaining Nut	1032
32	3BLA0162	1	Transmission Driveshaft Bearing Cap	2004
33	3BLA0452	1	Transmission Driveshaft Bearing Cap Gasket	1751
	S-15	2	Transmission Driveshaft Bearing Cap Screw-Upper	7A-503
	S-1782	2	Transmission Driveshaft Bearing Cap Screw-Lower	1971
	S-1347	4	Spring Washer	1A-600
			REVERSE GEAR GROUP	
34	3UK0270	1	Reverse Gear Assembly-with Bushing	50320
	3UK007	1	Reverse Gear Bushing	2735

Parts List No. 830
Bulletin Ref. #682

Sketch Ref. Number	Part No.	Req.	Description	Clark Equip. No. (Except as Noted)
			REVERSE GEAR GROUP (Continued)	
	3UK006	1	Reverse Gear Shaft	2719
	3BL01018	1	Reverse Gear Shaft Lock	1929
	S-50	1	Reverse Gear Shaft Lock Screw	7A-600
	S-1347	1	Spring Washer	1A-600
35	3UK001	1	Transmission Case-Bare	3179
	S-619	2	Drain & Filler Plug	11A-1200
	3BL0332	4	Transmission Case Clutch Housing Stud	1079
	S-84	4	Transmission Case Clutch Housing Stud Nut	10A-900
	S-1350	4	Spring Washer	1A-900
	3B3154	2	Power Take-off Opening Cover	1398
	3B3158	2	Power Take-off Opening Cover Gskt.	1400
	S-50	12	Power Take-off Opening Cover Screw	7A-600
	S-1347	12	Spring Washer	1A-600
			TRANSMISSION COVER, GEARSHIFT RODS AND FORK GROUP	
37	3UK016	1	Transmission Case Cover	2713
	S-23	6	Transmission Case Cover Screw	7A-608
	S-1347	6	Spring Washer	1A-600
38	3BL017	1	Transmission Case Cover Gasket	2733
39	S-4235	3	Gearshift Rod Hole Welch Plug	1420
40	3UK083	1	Third & Fourth Speed Gearshift Rod	2723
41	3UK01036	1	Third & Fourth Speed Gearshift Lug	2732
42	3UK015	1	Third & Fourth Speed Gearshift Fork	2736
43	3UK084	1	First & Second Speed Gearshift Rod	2722
44	3BL0191	1	First & Second Speed Gearshift Lug	1085
45	3UK087	1	First & Second Speed Gearshift Fork	2738
46	3UK085	1	Reverse Gearshift Rod	2721
47	3UB01035	1	Reverse Gearshift Lug	2949
	3UK035	1	Reverse Gearshift Fork	2724
	3BL0629	6	Gearshift Fork & Lug Screw	1566
48	3BL01007	1	Reverse Gearshift Latch Plunger	1042
49	3BL01008	1	Reverse Gearshift Latch Plunger Spring	1043
50	S-2024	1	Reverse Gearshift Latch Adjusting Nut	10A-500
51	S-1670	3	Gearshift Rod Plunger Ball	6A-600
52	3BL011	3	Gearshift Rod Spring	1844
53	3BL01009	2	Gearshift Rod Interlock Plug	1463

Parts List No. 830
Bulletin Ref. #682

Sketch Ref. Number	Part No.	Req.	Description	Clark Equip. No. (Except as Noted)
			TRANSMISSION COVER, GEARSHIFT RODS AND FORK GROUP (Continued)	
	3BL01011	1	Gearshift Rod Interlock Pin	1093
	S-4207	1	Gearshift Rod Interlock Hole Welch Plug	1423
			GEARSHIFT LEVER & BRACKET GROUP	
		1	Gearshift Lever Assembly	51094
65	10UK023	1	Gearshift Lever - Bare	4196
66	10B0408	1	Gearshift Lever Ball	2750
67	10UB01017	1	Gearshift Lever Dust Cover	2484
68	10UB0391	1	Gearshift Lever Pivot Spring	1298
69	10UB026	1	Gearshift Lever Pivot Washer	1299
70	10UB018	1	Gearshift Lever Mounting Bracket	2788
71	10UB0767	1	Gearshift Lever End	3036
72	S-1593	2	Gearshift Lever End Pin	3061
			GEARSHIFT CONNECTING ROD GROUP	
73	3UB01056	1	Gearshift Connecting Rod Yoke (Lever End)	2132
74	10UK01450	1	Gearshift Connecting Rod & Rear Yoke Assembly	51063
75	S-1457	2	Gearshift Connecting Rod Yoke Bolt	8A-602
	S-1418	2	Gearshift Connecting Rod Yoke Bolt Nut	10A-600
	S-1347	2	Spring Washer	1A-600
	3UB01055	6	Gearshift Connecting Rod Yoke Clevis Pin	2129
			GEARSHIFT TOP CONTROL COVER GROUP	
76	3UK0309	1	Gearshift Top Control Cover	2785
77	3UB0961	1	Gearshift Top Control Cover Gasket	2733
	S-23	4	Gearshift Top Control Cover Screw	7A-608
	S-1347	4	Spring Washer	1A-600
78	3UB01052	1	Pivot Anchor Bracket	2122
79	3UB01049	1	Pivot Anchor Bracket Snap Ring	2080
80	3UB088	1	Shift Lever Finger	2783
81	3UK01058	1	Shift Lever Rocker Shaft	2784
	S-4236	1	Shift Lever Rocker Shaft Hole Welch Plug	3070
82	S-2311	1	Shift Lever Finger Key	4A-11
	3UB01048	1	Shift Lever Finger Lock Screw	1705

-5-

Parts List No. 830
Bulletin Ref. #682

Sketch Ref. Number	Part No.	Req.	Description	Clark Equip. No. (Except as Noted)
			GEARSHIFT UNIVERSAL LEVER GROUP	
83	3UB01059	1	Gearshift Universal Lever	2936
84	3UB01053	1	Gearshift Universal Lever Bracket	2124
85	3UB01051	1	Gearshift Universal Lever Snap Ring	2081
86	S-4235	1	Gearshift Universal Lever Welch Plug	1420
87	3UB01054	1	Gearshift Universal Lever Boot	2128
88	3UB01057	1	Gearshift Universal Lever Finger	2934
	S-2311	1	Gearshift Universal Lever Finger Key	4A-11
	3UB01048	2	Gearshift Universal Lever Finger Lock Screw	1705
			SPEEDOMETER DRIVE GROUP	
89	16D0569	1	Speedometer Drive Gear	90059
90	3UK01370	1	Speedometer Drive Gear Case and Bushing Assembly	50360
91	3BLA0236	1	Speedometer Drive Gear Case Oil Seal	50035
	S-23	3	Speedometer Drive Gear Case Screw-Short	7A-608
92	S-40	1	Speedometer Drive Gear Case Screw-Long	7A-606
	S-1347	4	Spring Washer	1A-600
93	3UK0196	1	Speedometer Drive Gear Case Gasket	2716

SERVICE BULLETIN   No. 69

THE AUTOCAR COMPANY, ARDMORE, PENNSYLVANIA
SERVICE DEPARTMENT

SUBJECT   TIMKEN MODEL T-2-B7-4 TRANSFER CASE

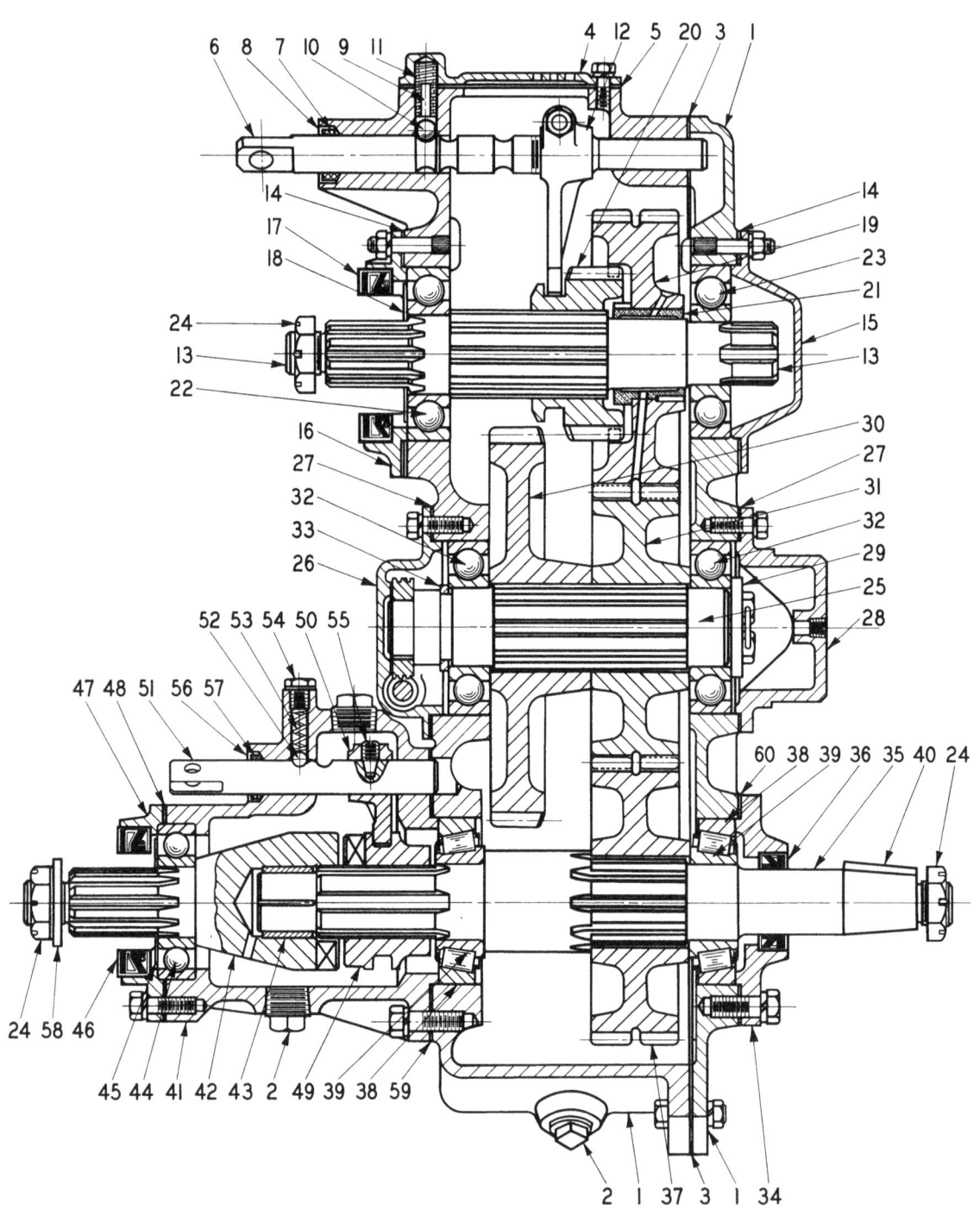

GENERAL

This unit which is used for the transfer of power to the front and rear axles on four wheel drive trucks is mounted to the rear of the main transmission and is arranged for two speeds, one in direct and the other in underdrive with a ratio of 1.95 to 1. Control for these speed changes is brought forward to a lever in the cab.

Page 32

These transfer cases are also equipped with a declutching device on the front end of the lower shaft and this device may be used to cut out the transmission of power to the front axle when traction on the front wheels is not required.

Control for this declutching device is mounted on the front end of the transfer case with a lock for the lever in either one of the two positions.

BEAR IN MIND THAT HEAVY LOADS SHOULD BE DIVIDED BETWEEN THE TWO AXLES AND THE TRUCK SHOULD DRIVE ON ALL FOUR WHEELS WHEN IN ANY KIND OF REAL HEAVY DUTY.

CONSTRUCTION

The construction is as shown in the illustration with the main gearshift rod at 6, the driveshaft from the engine at 13 and the two driveshafts to the front and rear axle at 42 and 35 respectively.

The fork for the declutching device is shown at 50 with the control shift rod at 51.

The main drive is through a set of helical herringbone type gears which provides quiet operation in all direct speeds.

All shafting is carried on annular ball bearings as illustrated.

LUBRICATION

Use S.A.E. No. 140 oil for Summer. Use S.A.E. No. 90 oil for Winter.

There is a drain plug in the bottom of the case, a level plug on the left side opposite the lower shaft and a breather further up on the left side of the case approximately level with the top of the cross member.

PARTS LIST No. 860

THE AUTOCAR COMPANY, ARDMORE, PENNSYLVANIA
SERVICE DEPARTMENT

BULLETIN REF. 697

ASSEMBLY — TIMKEN MODEL T-2-B7-4 TRANSFER CASE

Sketch Ref. Number	Part No.	Req.	Description	Timken Number
	19BK0440	1	Transfer Case Assembly (With Declutching Unit)	T-2-B7-4
			TRANSFER CASE GROUP	
1	19UG0590	1	Transfer Case & Cover Assembly	A3875-E-5
2	S-4131	3	Transfer Case Drain Plug	P-112
3	19DKB0185	1	Transfer Case Cover Gasket	2808-G-215
	19UG0619	20	Transfer Case Cover Screw	S-1613
	S-1418	24	Transfer Case Cover Screw Nut	N-16
	S-1347	46	Spring Washer	W-16
		4	Transfer Case Cover Pin	#7 x 1-1/4
			GEARSHIFT ROD & FORK GROUP	
4	19UG0188	1	Shifter Cover	3655-M-13
5	19SK0189	1	Shifter Cover Gasket	2808-H-216
	S-12	4	Shifter Cover Screw	S-266
6	19UG0191	1	Shifter Shaft	2843-U-99
7	19UG0623	1	Shifter Shaft Packing	5-X-287
8	19UG0625	1	Shifter Shaft Packing Retainer	2813-A-1
9	19SKB0463	1	Shifter Lock Plunger	1846-B-2
10	S-1671	1	Shifter Lock Ball	1898-E-57
11	19UG0202	1	Shifter Lock Plunger Spring	2858-D-82
12	19SK0205	1	Shifter Fork	2849-N-14
	19UG0627	1	Shifter Fork Screw	S-1614-X
	S-809	1	Shifter Fork Screw Nut	N-26
			MAIN DRIVE SHAFT GROUP	
13	19DKB0206	1	Main Drive Shaft	3880-H-60
14	19DKA0212	2	Main Drive Front & Rear Bearing Cover Gasket	2808-B-158
15	19DKA0211	1	Main Drive Shaft Rear Bearing Cover	3866-C-133
16	19DKB0570	2	Main Drive Front Cover	A3866-M-91
17	19DKB0214	1	Main Drive Front Oil Seal	A1805-B-80
18	19DKA0224	1	Main Drive Shaft Front Bearing Ring	1829-J-166
19	19UGA0225	1	Main Drive Shaft Direct Drive Gear (37-Teeth)	3894-K-167
20	19UG0225	1	Main Drive Shaft Slide Gear (21-T)	3892-V-282
21	19DKA0503	1	Main Drive Shaft Direct Drive Gear Bushing	1825-P-16

Page 34

Sketch Ref. Number	Part No.	Req.	Description	Timken Number
			MAIN DRIVE SHAFT GROUP (Continued)	
	19DKA0504	1	Main Drive Shaft Direct Drive Gear Bushing Pin	1846-S-19
22	3CKA2285	1	Main Drive Shaft Front Bearing	1309
23	19DKB0362	1	Main Drive Shaft Rear Bearing	1407
24	4A081	3	Main Drive & Driven Shaft Nut	13399
			IDLER SHAFT GROUP	
25	19DKB0215	1	Idler Shaft	3880-G-59
26	19DKA0216	1	Idler Shaft Front Bearing Cover	3866-D-108
27	19DKB0494	1	Idler Shaft Front & Rear Bearing Cover Gasket	2808-F-214
28	19DKA0131	1	Idler Shaft Rear Bearing Cover	3866-P-224
29	4BLA01178	1	Idler Shaft Rear Bearing Washer	1829-F-6
30	19UG0358	1	Idler Shaft Low Speed Gear (41-T)	3892-W-283
31	19DKB0357	1	Idler Shaft Driven Gear (37-T)	3894-R-122
32	3CKA2285	2	Idler Shaft Bearing	1309
33	19DKA0223	1	Idler Shaft Front Bearing Snap Ring	1854-S-19
			DRIVEN SHAFT GROUP	
34	4UG01430	1	Driven Shaft Rear Cover Assem.	A3866-D-342
35	19DKA0218	1	Driven Shaft	3880-P-42
36	19UG0214	1	Driven Shaft Oil Seal	A1805-M-169
37	19DKA0489	1	Driven Shaft Gear (37-Teeth)	3894-Q-121
38	4H2137	2	Driven Shaft Bearing Cup	432
39	4ZGE0653	2	Driven Shaft Bearing Cone	438
40	4NK0163	1	Driven Shaft End Key	16-X-28
			DECLUTCHING UNIT GROUP	
41	19UG0425	1	Declutching Shaft Carrier	3826-Q-173
42	19UG0427	1	Declutching Shaft & Bushing Assem.	A3880-N-40
43	19DKA0502	1	Declutching Shaft Bushing	1825-M-13
44	3SA2285	1	Declutching Shaft Bearing	1308
45	19DKA0429	1	Declutching Shaft Bearing Oil Slinger	1829-Y-103
46	19UG0431	1	Declutching Shaft Bearing & Cap Oil Seal	A1805-X-154
47	19UG0760	1	Declutching Shaft Bearing Cap Assembly	3866-L-64
48	19DKA0511	1	Declutching Shaft Bearing Cap Gasket	2808-J-88

Parts List No. 860
Bulletin Ref. #697

Sketch Ref. Number	Part No.	Req.	Description	Timken Number
			DECLUTCHING UNIT GROUP (Continued)	
49	19DKA0435	1	Sliding Clutch	2848-C-3
50	19UG0479	1	Declutching Shifter Fork	2849-G-85
51	19UG0437	1	Declutching Shifter Shaft	2843-P-94
52	S-1670	1	Declutching Shifter Shaft Lock Ball	1898-R-70
53	19UG0103	1	Declutching Shifter Shaft Lock Spring	2858-E-83
54	19UG0622	1	Declutching Shifter Lock Spring Screw	3-X-146
55		1	Declutching Shifter Lever Screw	26-X-52
56	19UG0626	1	Declutching Shifter Shaft Packing Retainer	2808-J-88
57	19UG0624	1	Declutching Shifter Shaft Packing	5-X-349
58	19DKA0513	1	Declutching Shaft End Washer	1829-K-115
		1	Declutching Carrier Oil Filler Plug	1850-Q-43
59	19DKA0512	1	Declutching Carrier Gasket	2808-U-281
59	19DKA0507	As	Declutching Carrier Shim-Thin	2803-P-224
59	19DKA0508	As	Declutching Carrier Shim-Medium	2803-Q-225
59	19DKA0509	As	Declutching Carrier Shim-Thick	2803-R-226
60	19DKA0479	1	Driven Shaft Rear Bearing Cover Gasket	2808-Q-17
60	19DKA0505	As	Driven Shaft Rear Bearing Cover Shim-Thin	2803-F-6
60	19DKA0506	As	Driven Shaft Rear Bearing Cover Shim-Thick	2803-Z-52

PARTS LIST No. 457X

THE AUTOCAR COMPANY, ARDMORE, PENNSYLVANIA
SERVICE DEPARTMENT   BULLETIN REF.

NEEDLE BEARING TYPE DRIVESHAFT AND COMPANION FLANGES
MODEL U-2044

ASSEMBLY

DRIVESHAFTS	PART NUMBER
Front Driveshaft	30ZC5330B
Intermediate Driveshaft	30LD4110
Rear Driveshaft	30DD5340

DRIVESHAFT FLANGES	
Main Transmission-Rear	3AH3201
Transfer Case-Upper Front	3DA3201
Transfer Case-Lower-Front	3BL3201A
Transfer Case-Lower-Rear	3DC3201
Rear Axle	3RG3201
Front Axle	3AM3201

SERVICE BULLETIN   No. 590

THE AUTOCAR COMPANY, ARDMORE, PENNSYLVANIA
SERVICE DEPARTMENT

SUBJECT   NEEDLE BEARING UNIVERSAL JOINTS - SPICER SERIES 1400 and 1410

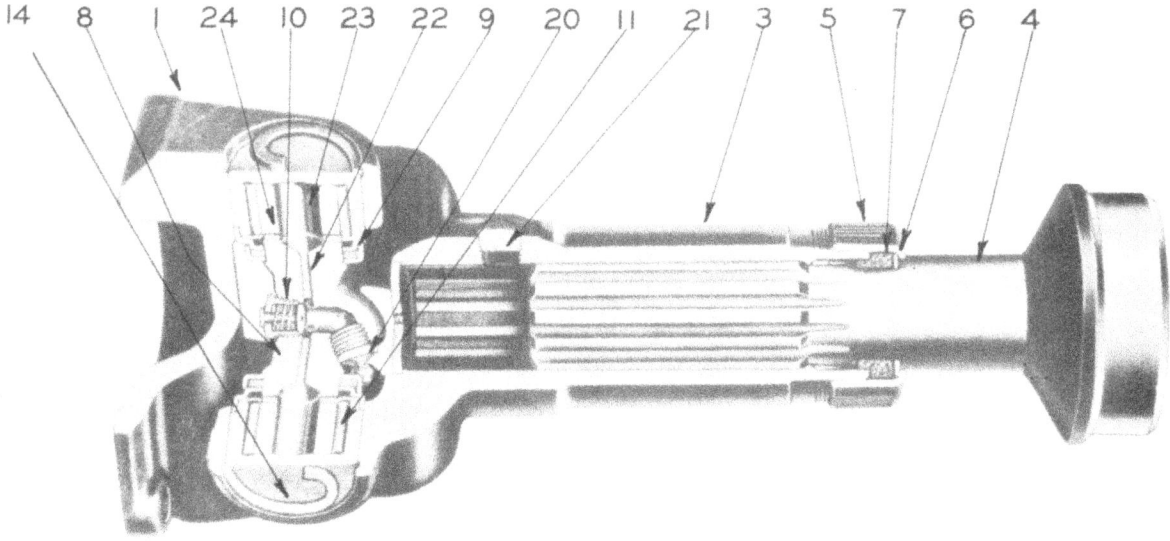

The Needle Bearing Universal Joint construction provides greater load capacity, less friction, greater efficiency, larger angular capacity, and much lower operating temperatures which preserve the consistency of the lubricant indefinitely and give longer life than will be experienced with corresponding sizes of any plain bearing joints.

LUBRICATION

DO NOT USE GREASE as this lubricant is liable to clog the oil passages. Use a heavy oil, SAE No. 90, for winter and SAE No. 140 for summer.

Lubricate universals and slip joint through fittings at 20 and 21 weekly or every 1000 miles of service.

From the high pressure fitting 20 the lubricant flows to a central chamber in the trunnion cross or journal 8, then through the drilled passages 22 to the four oil reservoirs 23 and through the small holes 24 to the needle bearing. A relief valve 10 is provided at the central chamber to prevent damage to the oil seals through high pressure fitting and to serve as an indicator when the joint is completely filled.

PARTS LIST

No. 839

THE AUTOCAR COMPANY, ARDMORE, PENNSYLVANIA
SERVICE DEPARTMENT

BULLETIN REF. 590

ASSEMBLY 30ZC5000B - NEEDLE BEARING TYPE DRIVESHAFT - 1410 SERIES

Sketch Ref. Number	Part No.	Req.	Description	Spicer Mfg. Co. No.
	3ZC0140A	1	Slip Joint Assembly (Flange Type with Outside Pilot)	8242-SF
	3DC0150A	1	Fixed Joint Assembly (Flange Type with Outside Pilot)	8128-SF
1	3DC0689	2	Joint Flange (With 2-3/4" Dia. Outside Pilot)	K3-2-159
3	3ZC0637	1	Slip Joint Sleeve Yoke	KL3-3-508X
4	3ZC0694A	1	Slip Spline Stub Shaft (3" Tube)	K3-40-341
	3DC0695	1	Fixed Stub Yoke (3" Tube)	K3-28-97
5	3DC0218	1	Sleeve Dust Cap	3-1/2-14-39
6	3DC0219	1	Sleeve Dust Cap Steel Washer	3-1/2-15-53
7	3DC0221	1	Sleeve Dust Cap Felt Washer	K3-16-53
8	3DC0632A	2	Journal Assembly	K3-5-108X
9	3DC0729	8	Journal Gasket	K3-86-89
	3DC0728	8	Journal Gasket Retainer	K3-76-17
10	3DD0755	1	Journal Relief Valve	98-798
11	3DC0635A	8	Needle Bearing Assembly	K3-6-68X
14	3DC0647	8	Needle Bearing Lock Ring	K3-7-39

SERVICE BULLETIN No. 397

THE AUTOCAR COMPANY, ARDMORE, PENNSYLVANIA
SERVICE DEPARTMENT

SUBJECT NEEDLE BEARING UNIVERSAL JOINTS - SPICER SERIES 1500-1600-1700

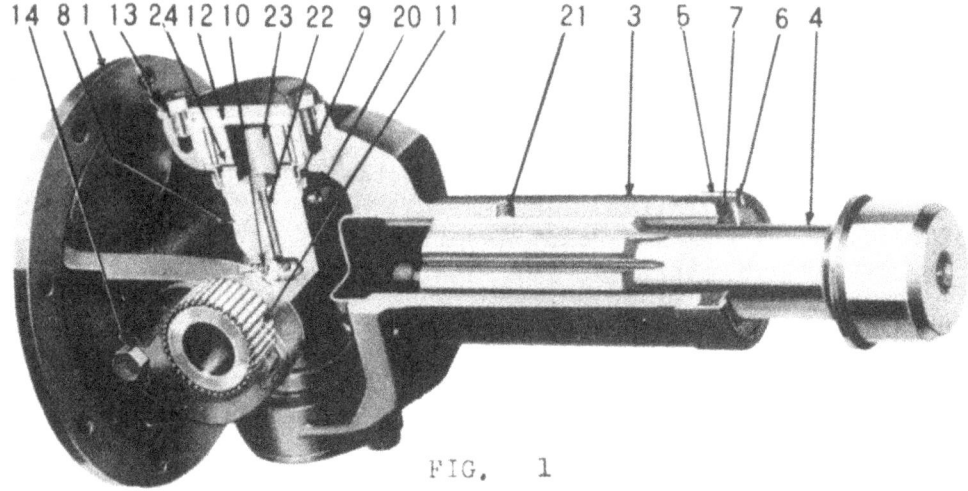

FIG. 1

The Needle Bearing Universal Joint construction provides greater load capacity, less friction, greater efficiency, larger angular capacity, and much lower operating temperatures which preserve the consistency of the lubricant indefinitely and give longer life than will be experienced with corresponding sizes of any plain bearing joints.

LUBRICATION

DO NOT USE GREASE as this lubricant is liable to clog the oil passages. Use a heavy oil, SAE No. 90 for winter and SAE No. 140 for summer.

Lubricate universals and slip joint through fittings at 20 and 21 weekly or every 1000 miles of service.

From the high pressure fitting 20 the lubricant flows to a central chamber in the trunnion cross or journal 8, then through the drilled passages 22 to the four oil reservoirs 23 and through the small holes 24 to the needle bearing. A relief valve 10 is provided at the central chamber to prevent damage to the oil seals through high pressure fitting and to serve as an indicator when the joint is completely filled.

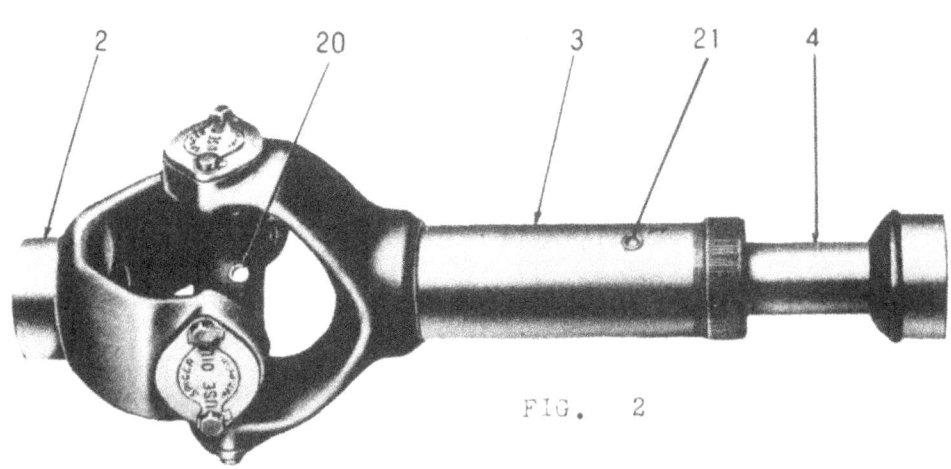

FIG. 2

Page 40

The bearings are protected against oil leakage or entrance of dirt by graphited oil seal 9.

SERVICE INFORMATION:

The journal and needle bearing assemblies are the only parts subject to wear and it is interesting to note that these parts can be replaced without special tools and without removing the driveshaft from the vehicle.

To disassemble, bend down locking lugs on plates holding screws 13, remove screws and plates 12 and 14. This permits removal of the needle bearing from the yoke by first tapping on the exposed face of one of the needle bearing cages to force out the opposite bearing assembly. Then tap the exposed end of the journal to free the second bearing assembly.

The journal can then be removed from the assembly by sliding toward one side of the yoke and lifting out at the opposite end.

To reassemble, merely reverse the above process. If slip joint assembly should be taken from the shaft for any reason, be sure when reassembling that the shaft yokes are located in the same plane in order to preserve the balance of the driveshaft assembly.

WHEN ASSEMBLING SLIP JOINT ON THE SHAFT, CARE MUST BE TAKEN TO ASSEMBLE SO THAT THE SLEEVE YOKE LUGS LIE IN THE SAME PLANE AS STUB BALL YOKE LUGS. ARROWS ARE STAMPED ON SLEEVE AND SHAFT FOR THIS PURPOSE.

PARTS LIST

No. 478

THE AUTOCAR COMPANY, ARDMORE, PENNSYLVANIA
SERVICE DEPARTMENT

BULLETIN REF. 397

ASSEMBLY 30LD4110 - NEEDLE BEARING TYPE DRIVESHAFT 1500 SERIES

Sketch Ref. Number	Part No.	Req.	Description	Spicer Mfg. Co. No.
	30LD4110	1	Driveshaft Assembly (11" long)	7035-SF
1	3ED0689	2	Joint Flange 5-3/4" O.D.	K4-2-299
3	3LD0637	1	Slip Joint Sleeve Yoke	K4-3-188X
4	3LD0753	1	Spline Stub Shaft	K4-82-61
5	3LD0218	1	Sleeve Dust Cap	K4-14-39
6	3DF0219	1	Sleeve Dust Cap Steel Washer	K5-15-23
7	3LD0221	1	Sleeve Dust Cap Felt Washer	K4-16-93
8	3RG0632	2	Journal Assembly	K4-5-78
9	3RG0729	8	Journal Gasket	K4-86-119
	3RG0728	8	Journal Gasket Retainer	K4-76-17
10	3DD0755	1	Journal Relief Valve	980798
11	3RG0635A	8	Needle Bearing Assembly	K4-6-68X
12	3RG0633	8	Needle Bearing Cap	K4-70-49
13	3GE0648	16	Needle Bearing Cap Screw	5-73-108
14	3RG0647	8	Needle Bearing Cap Screw Lock Plate	98-781

PARTS LIST

No. 464

THE AUTOCAR COMPANY, ARDMORE, PENNSYLVANIA
SERVICE DEPARTMENT

BULLETIN REF. 397

ASSEMBLY 30DD5000 - NEEDLE BEARING TYPE DRIVESHAFT - 1500 SERIES

Sketch Ref. Number	Part No.	Req.	Description	Spicer Mfg. Co. No.
	3DW0140	1	Slip Joint Assembly (Flange Type-Without Outside Pilot)	6908-SF
	3ED0150A	1	Fixed Joint Assembly (Flange Type-Without Outside Pilot)	7955-SF
1	3ED0689	2	Joint Flange 5-3/4" O.D.	K4-2-299
3	3RG0637	1	Slip Joint Sleeve Yoke	K4-3-88
4	3RG0694A	1	Slip Spline Stub Shaft (3" Tube)	K4-42-591
	3RG0695A	1	Fixed Stub Yoke (3" Tube)	K4-26-167
5	3RG0218	1	Sleeve Dust Cap	3-1/2-14-19
6	3RG0219	1	Sleeve Dust Cap Steel Washer	4-15-43
7	3RG0221	1	Sleeve Dust Cap Felt Washer	K4-16-83
8	3RG0632	2	Journal Assembly	K4-5-78
9	3RG0729	8	Journal Gasket	K4-86-119
	3RG0728	8	Journal Gasket Retainer	K4-76-17
10	3DD0755	1	Journal Relief Valve	98-798
11	3RG0635A	8	Needle Bearing Assembly	K4-6-68X
12	3RG0633	8	Needle Bearing Cap	K4-70-49
13	3GE0648	16	Needle Bearing Cap Screw	5-73-108
14	3RG0647	8	Needle Bearing Cap Screw Lock Plate	98-781

SERVICE BULLETIN No. 703

THE AUTOCAR COMPANY, ARDMORE, PENNSYLVANIA
SERVICE DEPARTMENT

SUBJECT DRIVESHAFT DISC BRAKE

The driveshaft disc brake which is operated by means of the hand brake lever should be used for emergency and parking only. It is a powerful, dependable brake, actuated by a lever mechanism providing positive release. The disc is ventilated to improve heat radiation and prevent overheating. Moulded brakeblok lining is used. This brake is very accessible so that it can be quickly and easily adjusted and brake shoes dismounted for relining within a few minutes' time.

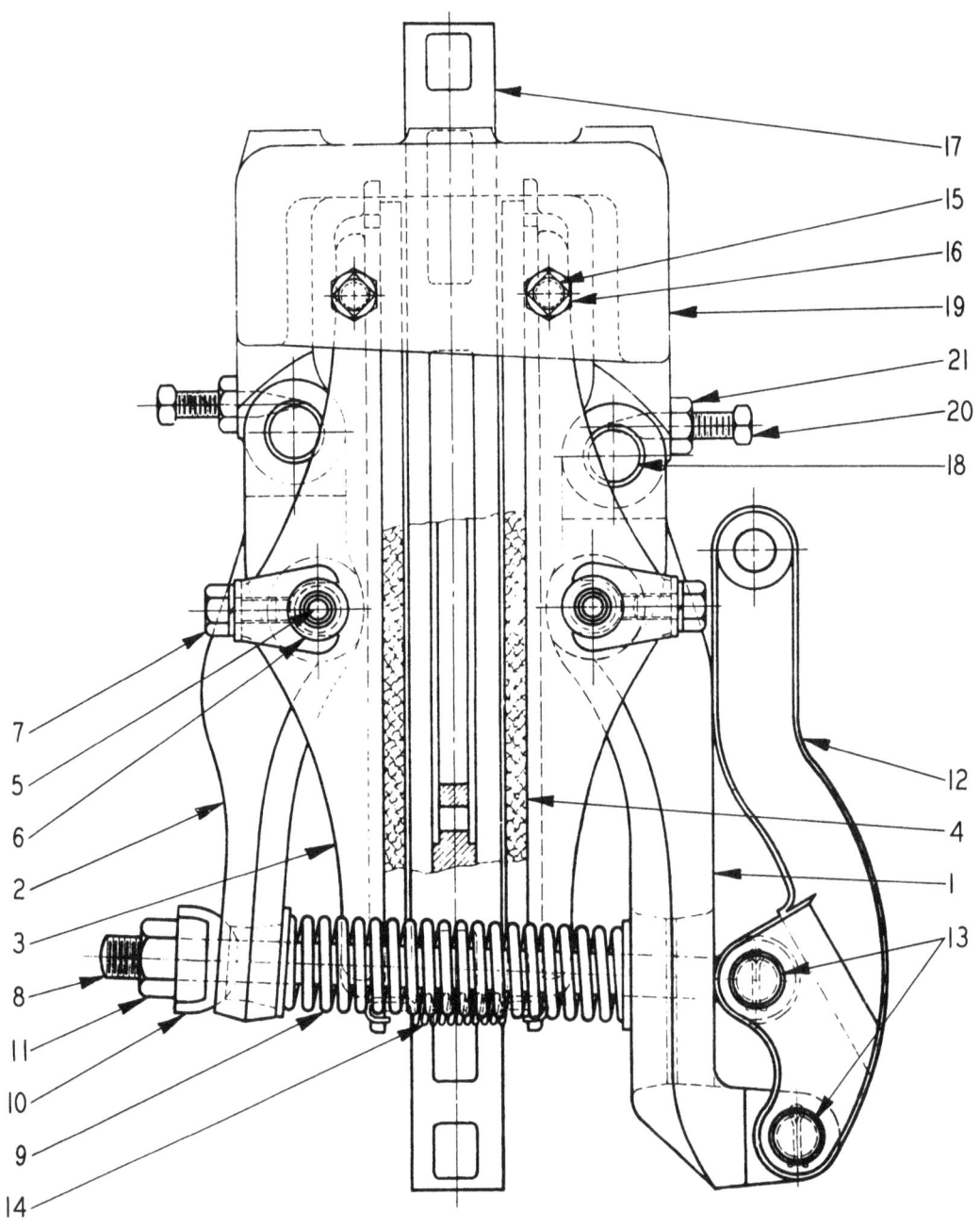

Page 44

OPERATION

This brake will ordinarily be severe in action and should be used with moderation for EMERGENCY and PARKING ONLY.

LUBRICATION

The only points requiring lubrication are the four brake shoe pins and the two brake cross shaft bearings. Lubricate weekly or every 1000 miles using heavy oil at the four brake shoe pins.

ADJUSTMENT

Tighten nut 10 until both brake shoes are solid against the disc. Then back off nut four full turns which will provide a clearance of 1/32" between the brake shoes and the disc. This clearance can also be checked by using 1/32" shims between the disc and brake shoes. When proper setting is obtained, tighten lock nut 11. See that tension spring 14 is in place and adjust screws 15 to set the shoe parallel with the brake disc. Proper adjustment of these screws can be made by running them in until the shoes are tight against the disc, then backing out 1-1/4 turns and setting lock nuts 16. Be sure to tighten lock nuts 16 on the brake shoe adjusting screws.

Never shorten the hand brake rod. Always make adjustments in accordance with the instructions given above.

SERVICE

To replace lining remove spring 14, lock nut 11 and adjusting nut 10 and apply hand brake lever so that the lever arm tie rod 8 can be withdrawn completely from the rear lever arm 2. The brake shoes can then be taken out by releasing the brake shoe pin locks 6 and removing the brake shoe pins 5.

PARTS LIST

No. 870

THE AUTOCAR COMPANY, ARDMORE, PENNSYLVANIA
SERVICE DEPARTMENT

BULLETIN REF. 703

ASSEMBLY 10HA7990 INSTALLATION - 14" LEVER TYPE DISC BRAKE

Sketch Ref. Number	Req.	Description	American Cable Co. No.
1	1	Lever Arm - Front L.H.	C-653
2	1	Lever Arm - Rear L.H.	C-655
	4	Bushing	C-598-2
3	2	Brake Shoe	C-506
4	2	Brake Shoe Lining	C-134
	16	Brake Shoe Rivet	C-241
	4	Bushing	C-598-4
5	2	Pin - Brake Shoe	C-509
6	2	Retainer - Brake Shoe Pin	C-542
7	2	Hex Head Cap Screw (5/16 x 18 x 5/8)	C-535
	2	Lockwasher (5/16 std.)	C-536
8	1	Tie Rod	C-479
9	1	Spring - Lever Arm Release	C-513
10	1	Spherical Nut	C-483-A
	2	Plain Washer	C-493
11	1	Nut (1/2-20 S.A.E. Hex)	C-270
12	1	Brake Operating Lever (L.H. Offset)	C-557
13	2	Clevis Pin (1/2" S.A.E. Std.)	RA-12-6
14	1	Brake Shoe Spring	C-514
15	2	Set Screw (Shoe Adjusting)	C-604
16	2	Locknut (3/8-16 U.S.S. Hex)	C-541
17		Brake Disc - 14"	40S-5D
18		Anchor Pin	C-507
		Bushing	C-598-2
19		Anchor Bracket	Autocar No. 10NC4995
20		Set Screw	Autocar No. 3Z282A
21		Lock Nut	Autocar No. S-3715

SERVICE BULLETIN No. 695

THE AUTOCAR COMPANY, ARDMORE, PENNSYLVANIA
SERVICE DEPARTMENT

SUBJECT   TIMKEN MODEL 5002-TW REAR AXLE

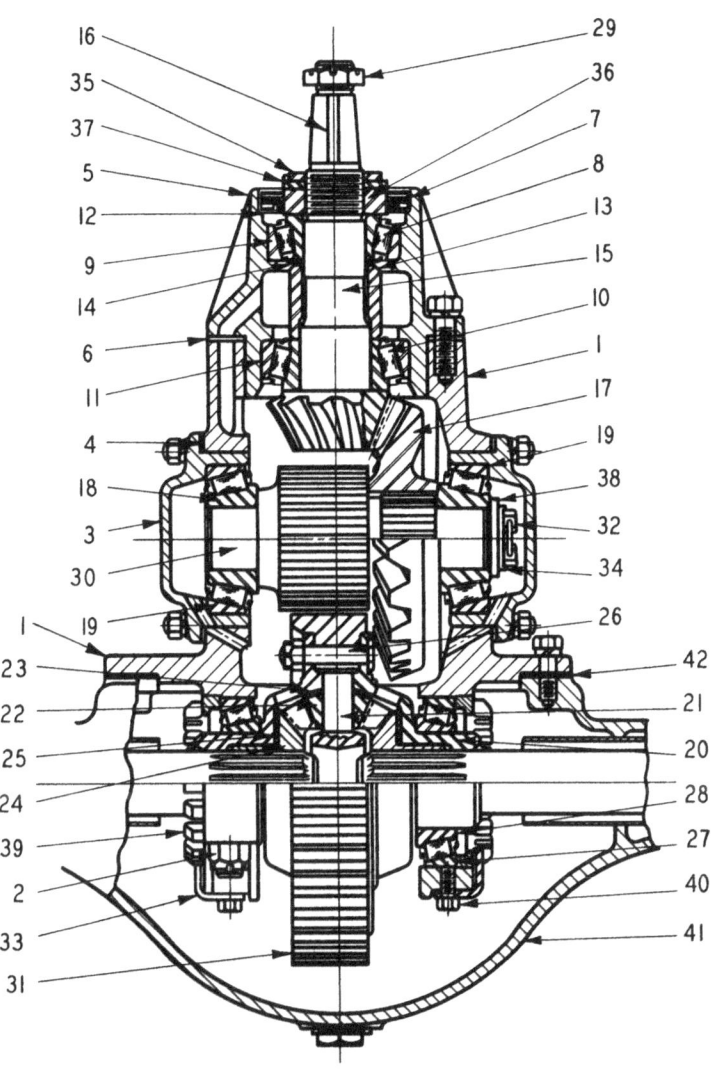

FIG. 1

GENERAL

The Timken Model 5002-TW rear axle is of the conventional double reduction design in which the first reduction is obtained through a set of spiral bevel gears and the second through a set of straight spur gears. The entire gearset is mounted in a gear carrier or cover plate which is detachable as a unit from the axle housing. The pinion, jackshaft and differential are mounted on tapered roller bearings. The axle housing is a large one piece steel casting with inserted tubes which carry the weight and provide for a full floating drive.

The gear carrier mechanism is illustrated in Fig. 1 and the wheel bearing and driving axle details are illustrated in Fig. 2.

LUBRICATION

Gear Carrier - Positive lubrication is provided for the pinion and jackshaft bearings by means of oil ducts which catch the throw-off from the bevel and ring gears while the differential unit runs directly in the lubricant reservoir.

Page 47

Bulletin #695

LUBRICATION - Gear Carrier - continued

Use SAE No. 140 oil for summer, SAE No. 90 oil for winter. DO NOT USE GREASE.

Check level of lubricant weekly or every 1,000 miles and add oil as required. Fill to level plug in rear of housing.

Every three months or every 10,000 miles, drain axle case, wash out with kerosene and refill with heavy oil to level plug.

Wheel Bearings - Lubricate wheel bearings through fittings in each wheel monthly or every 5,000 miles of service. Use U.S. Navy Grade No. 2 grease.

ADJUSTMENTS

Gear Carrier (Fig. 1) - The pinion bearings 8-10 which are of the tapered roller type are adjusted by means of shims 14 located between the outside bearing cone 8 and the spacer on the pinion shaft. Remove shims until bearings bind, then add a .001" or .002" shim to provide an adjustment of .000 - .002" tight.

To adjust mesh of pinion gear add or remove shims 6 under pinion cage. (For details on correct bevel gear adjustment refer to Service Bulletin #315).

Adjust jackshaft bearings 18 by removing or adding shims 4 under bearing caps at each end of the shaft. Remove shims to eliminate all play and adjust to .000 - .002" tight. Upon completion of the bearing set-up, check tooth contact of the bevel gear and pinion (Refer to Service Bulletin #315 for details on gear setting) and shift the jackshaft endways until correct tooth contact is obtained. In moving jackshaft, transpose shims 4 from one cap to the other so that bearing adjustment will not be changed.

The differential is mounted on Timken bearings 28 which are adjusted to .000" - .002" tight with adjuster ring 39 held in place with lock 33.

Rear Wheel Bearings (Fig. 2) - The wheel bearings 55-57 are Timken tapered roller type.

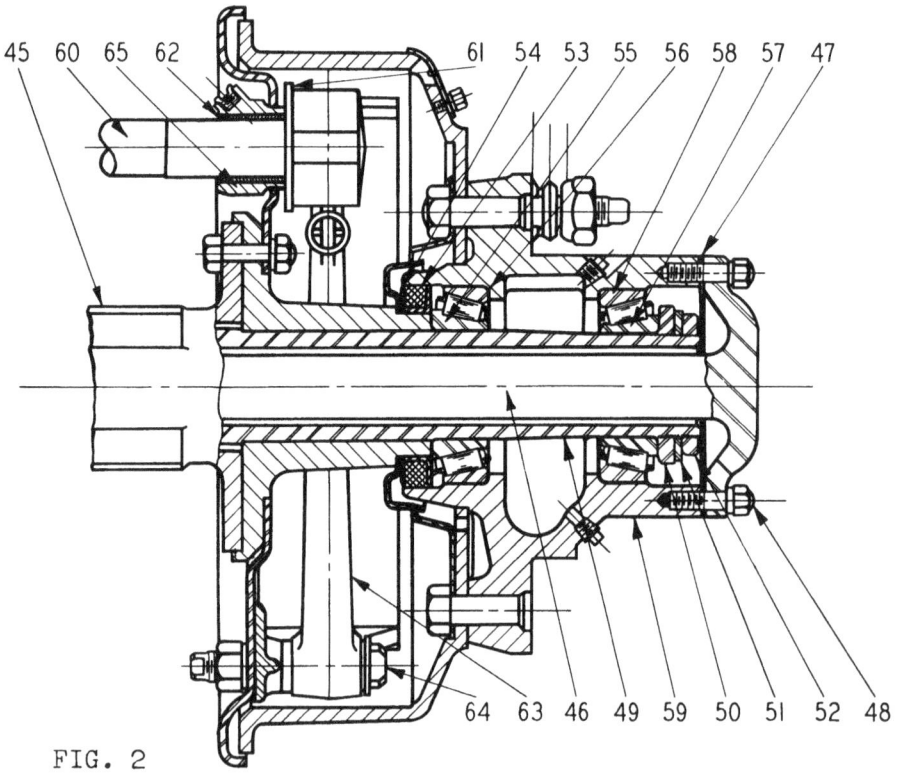

FIG. 2

ADJUSTMENTS - Rear Wheel Bearings - continued

To adjust, turn adjusting nut 50 against outer bearing until wheel binds, then back off nut 1/2 turn or until wheel rotates freely. Apply ring 51 and tighten jam nut 52. Bearings should be adjusted so that wheel will rotate freely without end play.

PARTS LIST

No. 847

THE AUTOCAR COMPANY, ARDMORE, PENNSYLVANIA
SERVICE DEPARTMENT

BULLETIN REF. 695

ASSEMBLY — TIMKEN MODEL 5002-TW-X1 REAR AXLE

Sketch Ref. Number	Part No.	Req.	Description	Timken Number
			GEAR CARRIER GROUP	
1	4BA0170	1	Mounting Assembly (Gear Carrier Assembly) Specify Ratio	G-316
1	4UG0920	1	Mounting with Caps (Gear Carrier)	A2-3800-K-63
	4NFG0122	4	Differential Support Cap Stud	4-X-177
2	S-810	4	Differential Support Strap Stud Nut	N-210
3	4BLA0460	2	Mounting Side Cap & Cup Assembly	A3838-Y-25
4	4BLA097	As	Mounting Side Cap Shim-Medium	2803-T-150
4	4BLA096	As	Mounting Side Cap Shim-Thick	2803-V-152
4	4BLA098	As	Mounting Side Cap Shim-Thin	2803-U-151
4	4BLA0143	2	Mounting Side Cap Gasket-R.H.	2808-F-58
	4NFG01084	14	Mounting Side Cap & Cage Stud	4-X-186
	S-1420	14	Mounting Side Cap & Cage Stud Nut	N-18
	S-4157	1	Mounting Inspection Plug	P-120
5	4UG040	1	Pinion Bearing Cage & Cup Assem.	A3826-Z-26
6	4BLA098A	As	Pinion Bearing Cage Shim-Thin	2803-Q-147
6	4BLA097A	As	Pinion Bearing Cage Shim-Medium	2803-R-148
6	4BLA096A	As	Pinion Bearing Cage Shim-Thick	2803-S-149
7	4UG084	1	Pinion Bearing Cage Oil Seal	A1805-F-162
8	4UG01232	1	Pinion Bearing Cone-Front	49175
9	4RL0137	1	Pinion Bearing Cup-Front	49368
10	4DFL0138	1	Pinion Bearing Cone-Rear	59200
11	4RL0142	1	Pinion Bearing Cup-Rear	59412
12	4UG0762	1	Pinion Bearing Spinner Ring	1829-U-21
13	4UG0584	1	Pinion Shaft Bearing Spacer	1844-W-23
14	4UG096	As	Pinion Shaft Bearing Shim	2803-L-142
14	4UGA096	As	Pinion Shaft Bearing Shim	2803-M-143
14	4UH096	As	Pinion Shaft Bearing Shim	2803-N-144
14	4UHA096	As	Pinion Shaft Bearing Shim	2803-P-146
15	4DK035	1	Bevel Pinion Shaft (10-T)	3890-V-74
16	4NK0163	1	Pinion Shaft Key	16-X-28
17	4BLA0521	1	Bevel Gear (21-T)	3889-H-60
18	4RL0141	2	Bevel Gear Bearing Cone	59175
19	4RL0142	2	Bevel Gear Bearing Cup	59412
	4UGA010	1	Differential & Ring Gear Assem. (Specify Ratio)	A3-3835-Y-51
	4UG010	1	Differential Assembly	A2-3835-Y-51
20	4BMC0220	1	Differential Cases Assem.	A3835-Z-52
21	4BLA058	1	Differential Spider	3878-B-54
22	4UG055	4	Differential Bevel Side Pinion	2233-M-429
23	4BMA0751	4	Differential Bevel Side Pinion Thrust Washer	1229-C-861

Parts List No. 847
Bulletin Ref. #695

Sketch Ref. Number	Part No.	Req.	Description	Timken Number
			GEAR CARRIER GROUP (Continued)	
24	4UG057	2	Differential Bevel Side Gear	2234-K-427
25	4BM0638	2	Differential Bevel Side Gear Thrust Washer	1229-Z-728
26	4BLA018	8	Differential Case Bolt	15-X-180
27	4D2157	2	Differential Bearing-Cup	472-A
28	19SKB0443	2	Differential Bearing-Cone	482
29	4A081	1	Bevel Pinion Shaft Nut	13399
30	4BLA044	1	Spur Pinion Shaft (8.21 Ratio)	3891-R-96
	4BMC01061	1	Spur Pinion Shaft (6.62 Ratio)	3891-Q-121
	4BMB062B	1	Spur Gear (Ring Gear) (8.21 Ratio)	3892-S-71
31	4BMC062	1	Spur Gear (Ring Gear) (6.62 Ratio)	3892-C-133
32	4BLA0648	2	Gear Shaft Bearing Washer Stud	4-X-322
33	4BMC0125	2	Differential Bearing Adjusting Nut Lock	1820-J-10
34	4BLA028	1	Gear Shaft Bearing Stud Lock	1820-K-11
35	4UG028	1	Pinion Bearing Lock	1827-G-7
36	4UG0936	1	Pinion Bearing Adjusting Nut	1827-N-14
37	4NK01204	1	Pinion Bearing Lock Washer	1829-E-5
38	4BLA01178	1	Gear Shaft Bearing Lock Washer	1829-F-6
	4NK0425	1	Carrier Plug	1850-B-28
39	4BMC0119	2	Differential Bearing Adj. Nut	2214-R-18
	S-811	8	Differential Case Bolt Nut	N-28
40	S-2065	2	Differential Bearing Adjusting Nut Lock Screw	S-265-D
	S-1349	14	Spring Washer	W-18
			REAR AXLE HOUSING GROUP	
45	4UG0660	1	Housing & Studs Assembly (Without Tubes)	A3801-Z-286
	4SK0417	1	Housing Filler Neck	3885-A-1
	4FD0176	1	Housing Filler Neck Plug	1850-A-1
		1	Housing Drain Plug	P-28
	4UG0483	1	Housing Gasket	2808-E-57
	4NFG01084	4	Carrier to Housing Stud	4-X-186
	S-830	5	Carrier to Housing Stud Nut	N-48
	S-19	8	Carrier to Housing Screw	S-2812
	S-1349	12	Spring Washer	W-18
	4BL01600	1	Housing Oil Breather	A1199-E-421
			REAR DRIVING AXLE & TUBE GROUP	
46	4DK038	2	Driving Axle	3802-A-53
47	4UG01143	2	Driving Axle Flange Gasket	2808-G-59
48	4UG087	24	Driving Axle Flange Stud	4-X-522
	4UG0730	2	Driving Axle Oil Seal	A1805-Y-103

Parts List No. 847
Bulletin Ref. #695

Sketch Ref. Number	Part No.	Req.	Description	Timken Number
			REAR DRIVING AXLE & TUBE GROUP (Continued)	
	4UG01165	2	Driving Axle Oil Seal Sleeve	2853-N-14
49	4AG022	2	Housing Tube	3816-J-36
			REAR HUB & WHEEL BEARING GROUP	
50	4D0550	2	Rear Hub Bearing Adjusting Nut	AT-6880
51	4D0166	2	Rear Hub Bearing Washer	T-3840
52	4D025	2	Rear Hub Bearing Adjusting Lock Nut	T-3564
	4DK0183	2	Rear Hub Bearing Felt Retainer	1805-Q-17
53	9DKA095	2	Rear Hub Bearing Felt Washer	5-X-282
54	4DK0107	2	Rear Hub Bearing Felt Washer Retainer	1829-F-214
55	4UG0141	2	Rear Hub Bearing-Cone-Inner	566
56	4Y2211	2	Rear Hub Bearing-Cup-Inner	563
57	4Y2141	2	Rear Hub Bearing-Cone-Outer	560
58	4Y2142	2	Rear Hub Bearing-Cup-Outer	552-A
59	4BP0811	2	Rear Hub & Cups Assembly	A333-E-395
	9UG0513	2	Rear Brake Drum Oil Slinger	3905-T-20
	4BP021	2	Rear Brake Drum	3819-R-44
	4A0499	20	Rear Budd Stud Nut	13332-E
	4A0503	10	Rear Wheel Budd Stud Nut-R.H.	37891-E
	4A0504	10	Rear Wheel Budd Stud Nut-L.H.	37892-E
	4UG0559	2	Rear Wheel Budd Spacer	3897-H-190
	4A0505	10	Rear Wheel Budd Stud Nut-R.H.	10708-E
	4A0506	10	Rear Wheel Budd Stud Nut-L.H.	10709-E
	4UK3489	10	Rear Wheel Budd Stud-R.H.	18309
	4UK3491	10	Rear Wheel Budd Stud-L.H.	18310
			REAR BRAKE GROUP	
60	4UG0251	1	Brake Shaft-R.H.	2810-R-252
60	4UG0252	1	Brake Shaft-L.H.	2810-Q-251
61	25G0106A	2	Brake Shaft Washer	R-5706
	25KC065	1	Brake Shaft Bracket & Bushing Assembly-R.H.	A3899-P-172
	25KD065	1	Brake Shaft Bracket & Bushing Assembly-L.H.	A3899-Q-173
62	4AKB0149	1	Brake Shaft Bracket Bushing-Thin	1825-P-42
63	4HDA0100	4	Brake Shoe Assembly	A3-3222-T-306
	4UG089	4	Brake Shoe Bushing	1225-U-99
	25SD0235	4	Brake Shoe Cam Plate	2217-C-26
	S-4280	4	Brake Shoe Cam Plate Screw	1846-X-1678
	4UG052	8	Brake Shoe Lining	2240-J-530

Parts List No. 847
Bulletin Ref. #695

Sketch Ref. Number	Part No.	Req.	Description	Timken Number
			REAR BRAKE GROUP (Continued)	
	S-1020	56	Brake Shoe Lining Rivet	X-1838
64	4HDA01317	4	Brake Shoe Anchor Pin	1259-M-39
	4HDA01078	4	Brake Shoe Anchor Pin Nut	13-X-22
	4HDA079	2	Brake Shoe Return Spring	2858-B-184
	S-563	12	Brake Anchor Bracket Rivet	X-240
	4NKA01205	2	Brake Spider to Housing Gasket	2808-L-168
	S-1438	8	Brake Spider to Housing Bolt	S-1816
	4HDA0254	2	Brake Dust Shield Assembly	A3236-H-502
	25AK0881	2	Brake Shaft Slack Adjuster	215530
	25KC021	1	Brake Air Chamber-R.H.	216475
	25KD021	1	Brake Air Chamber-L.H.	216476
	S-2536	4	Slack Adjuster Spacer Washer	1229-R-122
	4UG01353	2	Slack Adjuster Retainer Snap Ring	1854-B-80
	4UG0928	2	Brake Spider	3911-U-73
	25UG0393	4	Brake Air Chamber to Bracket Stud	4-X-541
	4SK0215	8	Brake Spider to Housing Bolt	3-X-135
65	4HDA0900	2	Brake Cam Shaft Support Bracket & Bushing Assembly	A3299-C-289
	25G0233A	2	Brake Cam Shaft Support Bracket Bushing	1225-H-86
	25G0103	4	Brake Shoe Anchor Pin "C" Washer	1229-M-221
	25N0112	4	Brake Cam Shaft Support Bracket Bolt	3-X-75
	S-1311	4	Brake Cam Shaft Support Bolt Nut	13-X-7
	S-1351	4	Spring Washer	X-533
	25G0243	4	Shakeproof Washer	1229-X-518

SERVICE BULLETIN No. 315

THE AUTOCAR COMPANY, ARDMORE, PENNSYLVANIA
SERVICE DEPARTMENT

SUBJECT   ADJUSTMENT OF SPIRAL BEVEL REAR AXLE GEARS

Proper adjustment of rear axle pinion and jackshaft bevel gear or bevel ring gear is absolutely essential to their proper operation both from the standpoint of noise and durability. Careful study and observance of the following simple instructions, and the cuts numbered Fig. 1 to Fig. 8 will enable any mechanic to perform this operation in a satisfactory manner.

The various gear tooth parts and terms referred to in adjustment procedure are detailed in Fig. 1 and Fig. 2.

For adjustment, remove the gear carrier from the axle and place in stand or vise where all gears and bearings are accessible.

Before proceeding with gear assembly, check the condition of the pinion, jackshaft and differential bearings, replacing any badly worn or pitted bearings and make sure that bearings are tight on shafts.

Mount pinion and bevel gear with back faces flush, set up bearings to recommended limits and adjust backlash in gears (See Fig. 2) between .005 and .015 as measured by an indicator, bearing on heel of gear tooth.

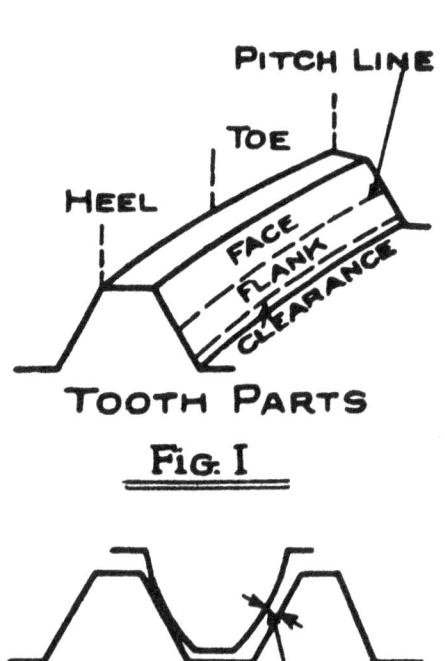

With the horizontal type carriers such as the SD and TF, the bevel gear adjustment should be made with the differential and gear removed from the carrier while on the vertical type such as the H and C, it will be necessary to assemble the arches and differential.

In the trial setups the pinion cage, jackshaft, caps or arches should be pulled tightly into place the same as in final assembly. Two bolts on opposite sides in the pinion cage and jackshaft caps will be sufficient for this purpose.

Now paint bevel gear teeth all around with a light coating of red lead, white lead or Prussian blue. Revolve gears and observe contact. You may find a condition of tooth contact shown on any of the figures from 3 to 8 on the following page.

CORRECT CONTACT UNDER LIGHT LOAD
Fig. 3

CORRECT CONTACT UNDER FULL LOAD
Fig. 4

HEEL CONTACT - TOO MUCH BACKLASH
Fig. 5

TOE CONTACT - TOO LITTLE BACKLASH
Fig. 6

NARROW HIGH CONTACT PINION TOO FAR OUT
Fig. 7

NARROW LOW CONTACT PINION TOO FAR IN
Fig. 8

FIGS. 3 and 4 both represent correct forms of tooth contact. It should be noted, however, that in connection with gears of some designs the correct contact will not cover the full width of the tooth face and there will be a strip along the top of the tooth from 1/32" to 1/16" wide which will show no contact. So long as the contact is of the approximate length and location shown in Figs. 3 and 4 and of fair width above and below pitch line, it may be considered satisfactory. At light load as turning the gear by hand a 3/4" to 7/8" contact starting at the toe (See Fig. 3) is to be desired. Under a heavier load as obtained by running the gears under load, a contact more closely approximating Fig. 4 should be obtained. However, any contact between these two conditions may be considered satisfactory. Continue to paint, run and adjust until proper contact is obtained.

FIG. 5 If the contact favors the heel as in Fig. 5, there is too much backlash. To remedy, adjust the gear toward the pinion.

FIG. 6 On the other hand, a contact favoring the toe as in Fig. 6, indicates too little backlash, which is corrected by adjusting the gear away from the pinion.

FIG. 7 If the contact is too narrow and toward the top of the tooth (as in Fig. 7), it indicates that the pinion is too far out and adjusting it in toward the gear will ordinarily widen out the bearing to approximate Figs. 3 and 4.

FIG. 8 is the exact opposite of the foregoing and may be corrected by moving the pinion out away from gear until correct contact is obtained.

In making pinion adjustments, be sure to check and maintain proper backlash. Moving the pinion back toward the gear reduces backlash and moving it forward away from the gear increases backlash. Adjust the differential or jackshaft sidewise to compensate for this loss or gain. The limits for backlash are from .005" to .015".

After all the foregoing adjustments have been made and the tooth contact approximates that shown in Figs. 3 and 4, see that all nuts, bolts, screws, etc., are securely locked and the parts free from dirt and grit. Replace rear axle carrier, fill axle to proper level with a good grade of axle lubricant, and road test car, listening for axle noise at all speeds from 5 to 40 miles per hour on both drive and coast. If there is noise, the condition may be improved by moving the pinion the thickness of one thin paper shim either way from the original position. This slight adjustment on the pinion will not materially change tooth contact and backlash.

PARTS LIST No. 871

THE AUTOCAR COMPANY, ARDMORE, PENNSYLVANIA
SERVICE DEPARTMENT   BULLETIN REF.

ASSEMBLY   16" x 3-1/2" AIR BRAKES ON TIMKEN 5002-TW-X1 REAR AXLE

Part No.	Req.	Description	Timken Number
4UG0251	1	Brake Shaft-R.H.	2810-R-252
4UG0252	1	Brake Shaft-L.H.	2810-Q-251
25G0106A	2	Brake Shaft Washer	R-5706
25KC065	1	Brake Shaft Bracket & Bushing Assembly-R.H.	A3899-P-172
25KD065	1	Brake Shaft Bracket & Bushing Assembly-L.H.	A3899-Q-173
4AKB0149	1	Brake Shaft Bracket Bushing-Thin	1825-P-42
4HDA0100	4	Brake Shoe Assembly	A3-3222-T-306
4UG089	4	Brake Shoe Bushing	1225-U-99
25SD0235	4	Brake Shoe Cam Plate	2217-C-26
S-4280	4	Brake Shoe Cam Plate Screw	1846-X-1678
4UG052	8	Brake Shoe Lining	2240-J-530
S-1020	56	Brake Shoe Lining Rivet	X-1838
4HDA01317	4	Brake Shoe Anchor Pin	1259-M-39
4HDA01078	4	Brake Shoe Anchor Pin Nut	13-X-22
4HDA079	2	Brake Shoe Return Spring	2858-B-184
S-563	12	Brake Anchor Bracket Rivet	X-240
4NKA01205	2	Brake Spider to Housing Gasket	2808-L-168
S-1438	8	Brake Spider to Housing Bolt	S-1816
4HDA0254	2	Brake Dust Shield Assembly	A3236-H-502
25AK0881	2	Brake Shaft Slack Adjuster	215530
25KC021	1	Brake Air Chamber-R.H.	216475
25KD021	1	Brake Air Chamber-L.H.	216476
S-2536	4	Slack Adjuster Spacer Washer	1229-R-122
4UG01353	2	Slack Adjuster Retainer Snap Ring	1854-B-80
4UG0928	2	Brake Spider	3911-U-73
25UG0393	4	Brake Air Chamber to Bracket Stud	4-X-541
4SK0215	8	Brake Spider to Housing Bolt	3-X-135
4HDA0900	2	Brake Cam Shaft Support Bracket & Bushing Assembly	A3299-C-289
25G0233A	2	Brake Cam Shaft Support Bracket Bushing	1225-H-86
25G0103	4	Brake Shoe Anchor Pin "C" Washer	1229-M-221
25N0112	4	Brake Cam Shaft Support Bracket Bolt	3-X-75
S-1311	4	Brake Cam Shaft Support Bolt Nut	13-X-7
S-1351	4	Spring Washer	X-533
25G0243	4	Shakeproof Washer	1229-X-518
4BP021	2	Rear Brake Drum	3819-R-44

PARTS LIST No. 864

THE AUTOCAR COMPANY, ARDMORE, PENNSYLVANIA
SERVICE DEPARTMENT BULLETIN REF.

ASSEMBLY: REAR SHOCK ABSORBERS - U-2044 & U-4044 - U.S.A. 1940

Part No.	Req.	Description
28M201	4	Shock Absorber Stud (Top)
28MA201	4	Shock Absorber Stud (Bottom)
S-77	16	1/2"-13 Hex Nut
S-1349	8	1/2" Spring Washer
S-390	8	Washer (17/32" I.D. x 1-3/8" O.D. x 1/16" Thk.)
S-2198	8	3/32 Cotter Pin
28M205	8	Shock Absorber Stud Spacer
28MA010	4	Shock Absorber
28UK211	2	Shock Absorber Frame Bracket (R.H.)
28UK212	2	Shock Absorber Frame Bracket (L.H.)
S-1872	8	1/2"-13 Hex Cap Screw 1-1/2" Long
S-77	8	1/2"-13 Hex Nut
S-1349	16	1/2" Spring Washer
13UG3113	1	Rear Spring Shim
13UG3114	1	Rear Spring Shim

PARTS LIST

No. 855

THE AUTOCAR COMPANY, ARDMORE, PENNSYLVANIA
SERVICE DEPARTMENT

BULLETIN REF.

ASSEMBLY: COOLING SYSTEM - MODEL U-2044 - U.S.A. 1940

Part No.	Req.	Description
5UK490	1	Radiator Core and Tanks
5UBL2300	1	Radiator Filler Cap
5BL2369	1	Radiator Drain Cock
5Y305	1	Radiator Outlet Elbow
5UK462	1	Water Outlet Elbow
5BL3158	1	Water Pump Inlet Elbow
5UK461	1	Radiator Inlet Elbow
S-2365	1	1-1/2" Rubber Hose - 8" Long (Engine to Radiator)
S-2365	1	1-1/2" Rubber Hose - 14-1/2" Long (Radiator to Pump)
5BL217	1	Brass Tube - 1-1/2" Outside Diameter - 6-1/4" Long (Locate inside of S-2365)
S-2404	4	Hose Clamp
5A230	1	Thermostat
5Y156	2	Cab Support Pad (Thermoid)
5ZBL256	2	Radiator Support Pad (Rubber)
5UK0104	2	Radiator Support Bracket
5UK4120	1	Fan and Hub Assembly
5UK088	1	Fan - 21" with 6 Blades
5UK355	1	Fan Belt
5UBA054	1	Fan Bracket
5RLC2381	1	Radiator Shell Weather Strip (142 inches)
5UBK2374	1	Radiator Top Support
5URK4540	1	Radiator Grille Assembly
5UK4553	1	Radiator Cross Member Assembly

SERVICE BULLETIN   No. 623

THE AUTOCAR COMPANY, ARDMORE, PENNSYLVANIA
SERVICE DEPARTMENT

SUBJECT   WATER PUMP FOR HERCULES JX ENGINES

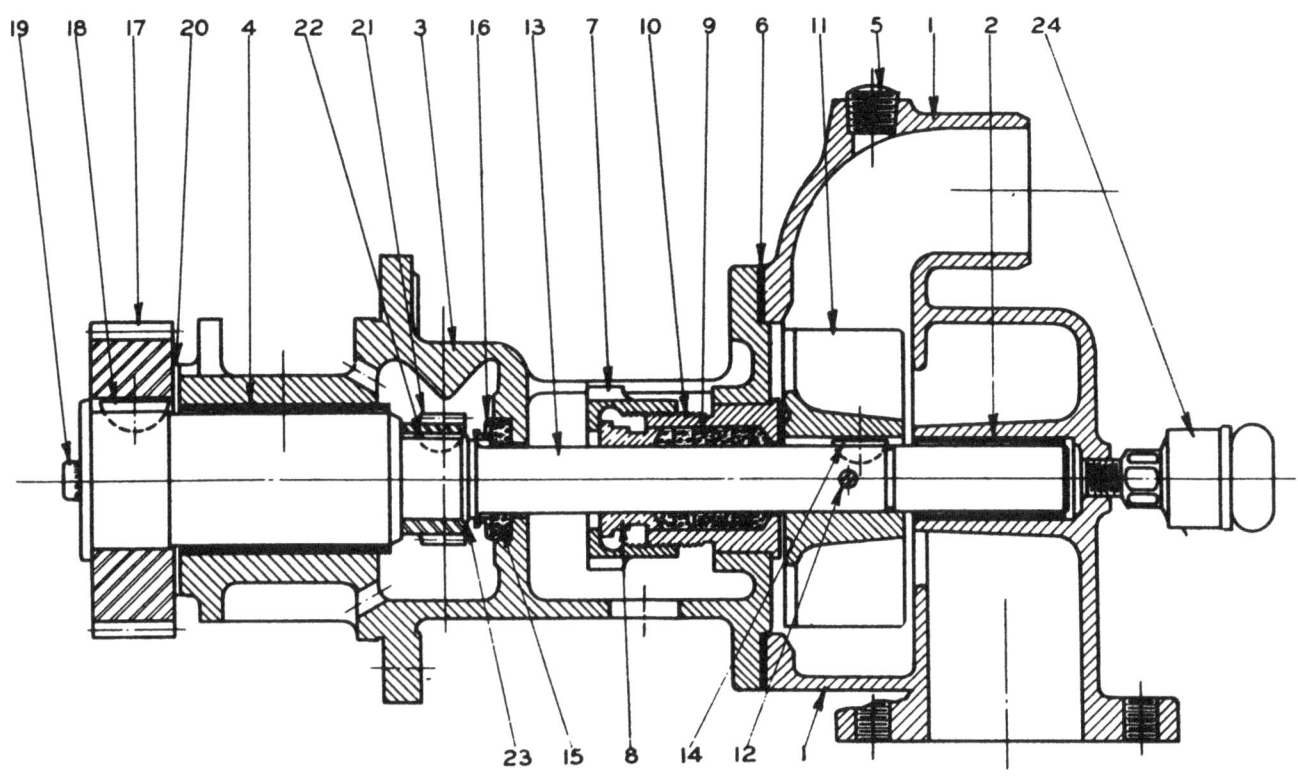

CONSTRUCTION

The water pump is mounted on the left hand side of the engine on an extension of the accessory driveshaft. The pump is of the centrifugal type with a single packing gland at the driven end. The entire water pump and accessory shaft assembly is attached to the crankcase at the timing gear case rear flange and may be removed from the engine by disconnecting at this flange and removing hose connections.

PACKING GLAND

The packing 9 is located in bushing 10 pressed in the rear end of the water pump head 3 and is held in place by the packing gland sleeve 8 and packing nut 7.

When tightening packing nut use sufficient force to compress the packing slightly but DO NOT TIGHTEN EXCESSIVELY. As scoring of the shaft or other damage may result. A special spanner wrench, part number X-23072 is available for this purpose.

If tightening packing nut does not stop leakage, remove nut and sleeve and install new packing. Split ring packing is furnished for service to facilitate replacement.

LUBRICATION

Screw down grease cup 24 on the rear end of the pump one or two turns weekly or every 1000 miles of service. Use a tallow base grease or other lubricant which will stand contact with hot water.

PARTS LIST

No. 534

THE AUTOCAR COMPANY, ARDMORE, PENNSYLVANIA
SERVICE DEPARTMENT

BULLETIN REF. 623

ASSEMBLY   WATER PUMP FOR HERCULES JX ENGINES - MODEL U-2044 CHASSIS

Sketch Ref. Number	Part No.	Req.	Description	Hercules Motors Corp. No.
	5BL040	1	Water Pump Assembly	40170-CS
1	5BL001	1	Water Pump Body & Bushing Assy.	45735-CS
2	5BL013	1	Water Pump Body Bushing	40168-A
3	5BL002	1	Water Pump Head & Bushing Assy.	40171-C
4	5BL086	1	Water Pump Head Bushing	22171-B
	S-23	3	Water Pump Attaching Screw	1864-A
	S-1347	3	Spring Washer	342-A
5	S-629	1	Water Pump Head Pipe Plug	305-A
	S-16	4	Water Pump Head Screw	303-A
	S-1346	4	Spring Washer	615-A
6	5BL004	1	Water Pump Gasket	22164-A
7	5BL032	1	Water Pump Packing Nut	22177-A
8	5BL081	1	Water Pump Packing Gland	21265-A
9	5BL0209	6	Water Pump Packing	22166-A
10	5BL087	1	Water Pump Packing Bushing	22175-A
11	5BL003	1	Water Pump Paddle	40165-A
12	S-1558	1	Water Pump Paddle Pin	14507-A
13	5BL012	1	Water Pump Paddle Shaft	40172-B
14	S-2305	1	Water Pump Paddle Key	513-A
15	5BL0393	1	Water Pump Shaft Cork Washer	4661-A
16	5BL0395	1	Water Pump Shaft Cork Washer Retainer	23116-A
17	2BL091	1	Water Pump Driving Gear	22195-B
18	S-2319	1	Water Pump Driving Gear Key	4265-A
19	2BL0776	1	Water Pump Thrust Plunger	40068-A
20	5BL0392	1	Water Pump Thrust Washer	4024-A
21	2BL0589	1	Water Pump Distributor Driving Gear	22336-A
22	S-2325	1	Water Pump Distributor Driving Gear Key	1179-A
23	2BL01209	1	Water Pump Distributor Driving Gear Key Snap Ring	22168-A
24	5TH0238	1	Grease Cup	749-A
	5BL0246	1	Water Pump Outlet Hose	40064-A
	5TH0247	2	Water Pump Outlet Hose Clamp	8159-A
	5BL3158	1	Water Pump Inlet Elbow	
	S-64	2	Water Pump Inlet Elbow Screw	
	S-1347	2	Spring Washer	
	S-1478	1	Water Pump Inlet Elbow Gasket	
	5BL2369	1	Water Pump Inlet Elbow Pet Cock	6202
	2BL0467	1	Water Pump Attaching Gasket	22149-A
	8BL0529	1	Distributor Arm Attaching Bracket	4343-A
	S-1879	1	Distributor Arm Attaching Bracket Screw	4266
	S-1347	1	Spring Washer	342
	X-23072	1	Water Pump Spanner Wrench	

SERVICE BULLETIN   No. 663

THE AUTOCAR COMPANY, ARDMORE, PENNSYLVANIA
SERVICE DEPARTMENT

SUBJECT   FAN DRIVE ASSEMBLY   -   MODEL U-2044

CONSTRUCTION

The fan assembly (see illustration) is attached to a bracket mounted on the timing gear case at front end of the engine and is driven by a V-belt from the engine crankshaft. This assembly is equipped with Timken tapered roller bearings which are adjustable for wear.

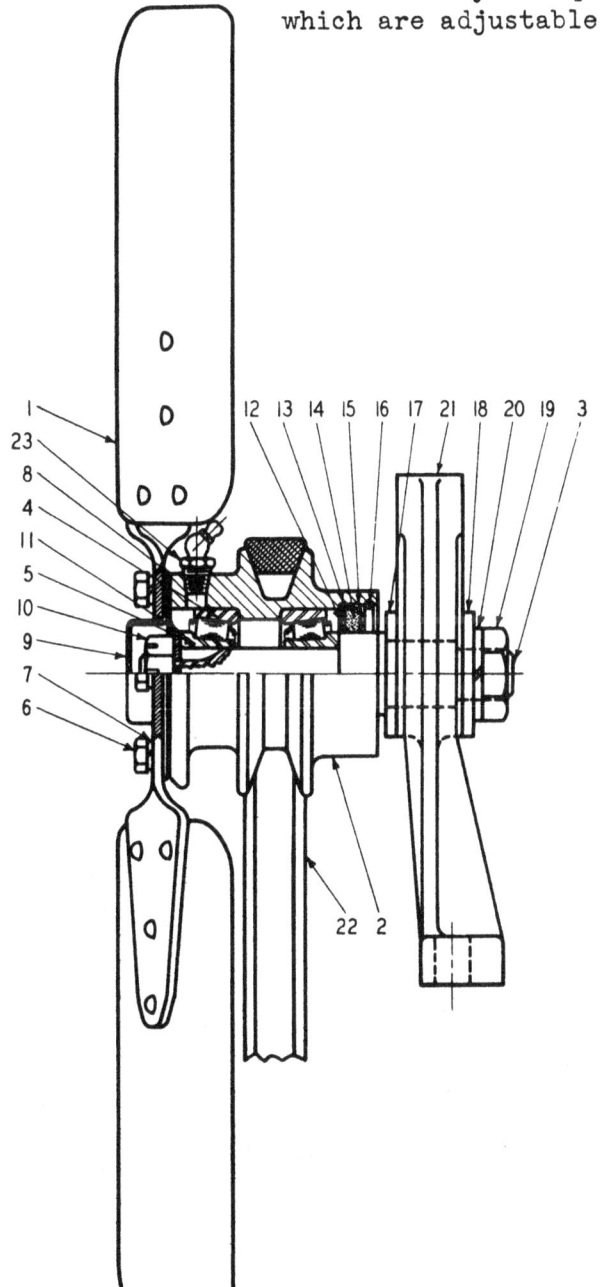

LUBRICATION

The fan hub which forms the housing for the two bearings contains a reservoir which is filled through the high pressure fitting 23. Lubricate assembly weekly or every 1000 miles using SAE No. 90 for winter and SAE No. 140 for summer. DO NOT OVERFILL AS GREASE MAY BE FORCED THROUGH OIL SEAL ON TO FAN PULLEY AND RESULT IN BELT SLIPPAGE.

BELT ADJUSTMENT

Fan belt adjustment is obtained by loosening the unit 19 at the rear of the fan shaft and moving the assembly up and down in the slotted bracket 21. ADJUSTMENT SHOULD BE SUFFICIENTLY TIGHT TO GIVE BELT A FIRM GRIP ON THE PULLEYS BUT NOT TIGHT ENOUGH TO STRETCH AND DAMAGE BELT. With proper adjustment, a 5-7 pound pull on the end of fan blade will be required to rotate fan.

FAN BEARINGS ADJUSTMENT

To adjust fan shaft bearings take out four screws 6, remove fan 1 from hub 2 and adjust nut 10. Adjustment should eliminate end play but permit assembly to rotate freely on bearings. Oil seal 14 should be inspected and replaced if worn.

PARTS LIST

No. 836

THE AUTOCAR COMPANY, ARDMORE, PENNSYLVANIA
SERVICE DEPARTMENT

BULLETIN REF. 663

ASSEMBLY 21" DIA. FAN ASSEMBLY CAB-OVER-ENGINE MODELS WITH JXD ENGINE

Sketch Ref. Number	Part No.	Req.	Description	Schwitzer Cummins
	5UK4120	1	Fan & Hub Assembly-21" with 6 Blades	
1	5UK088	1	Fan - 21" with 6 Blades	BF-06921
2	5ZBLE049A	1	Fan Hub & Pulley	C-101572
3	5ZBLE051A	1	Fan Spindle	C-101573
4	5SCM0256	2	Fan Spindle Bearing-Cup	*09194
5	5SCM048	2	Fan Spindle Bearing-Cone	*09074
6		4	Fan Screw	C-11922
7	S-1346	4	Spring Washer	C-594
8	5A022	2	Fan Hub Gasket	C-2016
10	5SCM047	1	Fan Spindle Nut-Front	C-2750
11	5TL072	1	Fan Spindle Nut Washer-Front	C-17065
12	5SCM0242	1	Fan Bearing Oil Gasket	C-4391
13	5SA073	1	Fan Bearing Cork Retaining Washer	C-2389
14	5SA074	1	Fan Bearing Cork Washer	C-3814
15	5SA053	1	Fan Bearing Cork Retainer	C-2308
16	5SCM071	1	Fan Bearing Cork Retainer Lock Wire	C-5098
17	5SA0274	1	Fan Support Bracket Clamp Washer-Front	C-2662
18	5SA0275	1	Fan Support Bracket Clamp Washer-Rear	C-2736
19	5A0192	1	Fan Support Bracket Clamp Nut	C-2673
20		1	Fan Support Bracket Cotter Pin	
21	5ZBLE054A	1	Fan Support Bracket	A-11239
		2	Fan Support Bracket Screw	**8533A
		2	Spring Washer	**312A
22	5UK055	1	Fan Belt	**42255B
		1	Fan Drive Pulley	**40500B
		1	Fan Drive Pulley Key	**4265A
		1	Fan Blade Spacer	C-110655
	5ZBLE007	1	Fan Adjusting Screw	C-15295
	5A047	1	Fan Adjusting Screw Nut	C-2692
	S-1349	1	Spring Washer	C-2675

*Timken Numbers

**Hercules Numbers

PARTS LIST

No. 853

THE AUTOCAR COMPANY, ARDMORE, PENNSYLVANIA
SERVICE DEPARTMENT

BULLETIN REF.

ASSEMBLY FUEL SYSTEM - MODEL U-2044 - U.S.A. 1940

Part No.	Req.	Description
6URK4720	1	Gas Tank Assembly
6UBK3610	1	Gas Tank Bracket - Front
6UBKA3610	1	Gas Tank Bracket - Rear
6URK215	2	Gas Tank Strap
6URK0730	1	Gas Tank Cap and Chain
16DFLB3786-15	1	Gas Tank Gauge
6UG0510	1	Fuel Pump
6T2502A	1	Fuel Pump Gasket
6UG450	1	Carburetor
6BL2375	2	Carburetor Gasket
6ZBM0330	1	Air Cleaner
6UBK3686	1	Air Cleaner Bracket
6UBK244	1	Air Cleaner Pipe
6UBK2239	1	Air Cleaner Pipe Adapter Elbow
6T3401	1	Air Cleaner Pipe Elbow
6URK3559	1	Air Cleaner Cover
6URK2111	1	Air Cleaner Cover Support

GASOLINE LINE & CONNECTIONS - TANK TO PUMP

Part No.	Req.	Description
6D022	1	1/4" Gas Shut-off Valve
S-5066	1	3/8" Single Elbow
S-5005	2	3/8" Union Nut
S-5146-40	1	3/8" Copper Tubing - 40" Long
S-4617	1	1/4" Service Elbow
6D3740	1	Titeflex Line (To Pump)

GASOLINE LINE & CONNECTIONS - PUMP TO CARBURETOR

Part No.	Req.	Description
S-5024	1	5/16" Single Union
S-5024A	1	5/16" Single Union
S-5013	2	5/16" Union Nut
S-5150-54	1	5/16" Copper Tubing - 54" Long
16Y0273	1	Loom

SERVICE BULLETIN No. 691

THE AUTOCAR COMPANY, ARDMORE, PENNSYLVANIA
SERVICE DEPARTMENT

SUBJECT CARBURETOR - ZENITH 457-2 FOR HERCULES JXD ENGINE

CONSTRUCTION

The Model 457-2 Carburetor is of the plain tube type with an adjustable main jet, an accelerating pump and an economizing device.

The main jet determines the maximum amount of fuel which may be obtained for high speed operations. The main jet adjustment reduces this amount if it is turned toward its seat. Ordinarily the main jet adjustment has no effect after it is two turns open.

To set this adjustment, retard the spark and open the throttle to approximately 1/4 open. Turn the adjustment clockwise, shutting off the fuel until the engine speed decreases due to too lean mixture. Now open the adjustment until the engine speed decreases due to too much fuel. The adjustment should be set at a position half way between these two extremes.

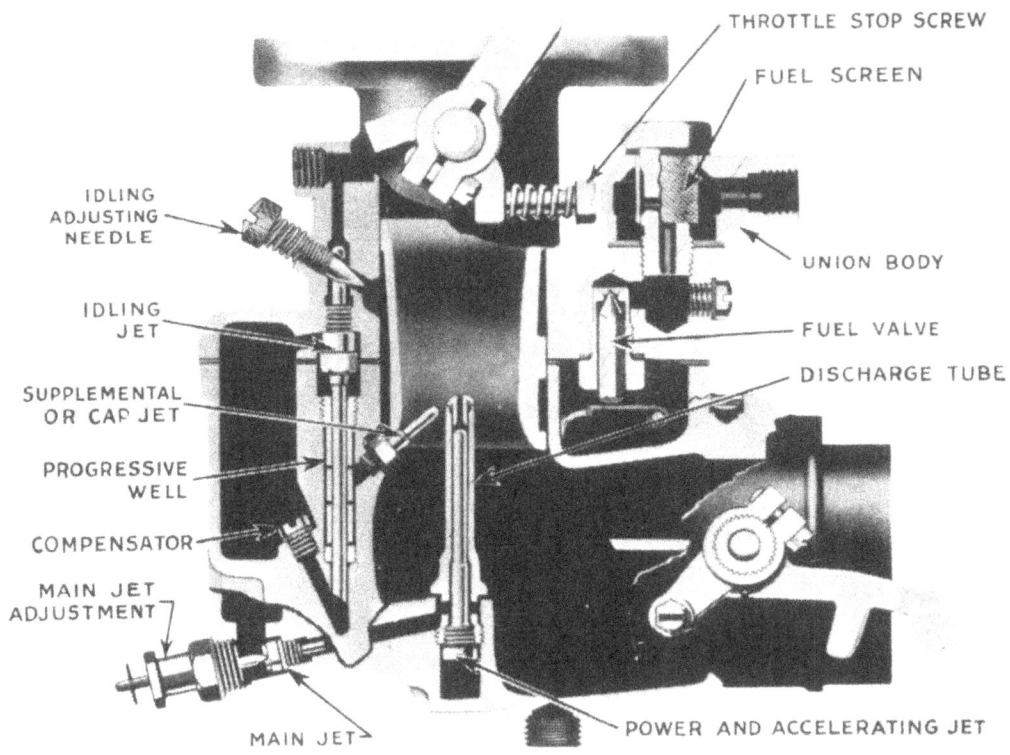

Figure 1

OPERATION

The Zenith Compound Nozzle System of carburetion is used in this model. This consists of two jets--the Main Jet, directly connecting fuel in the bowl with the air stream in the carburetor barrel through the Main Jet Discharge Tube; and the Compensating Jet flowing into an open well and connected with the air stream through the Supplemental Jet.

The Main Jet flow varies with suction, delivering more fuel as the engine speed increases, thus its tendency is to richness at top engine speed. The Compensating Jet is not affected by suction, thus flows the same at all speeds and has a tendency to leanness at top engine speed. In combination, the rich and the lean jets give an average mixture of correct proportion.

IDLING

The idling system functions only on starting and idling. When the throttle is opened past the idling position, the fuel goes the other way through the discharge tubes and the idling system is automatically out of operation.

It consists of an Idling Jet and tube to supply the fuel, an Idling Needle Valve to correct the idling mixture, and a channel to carry the mixture into the carburetor barrel at the edge of the throttle.

The desired idling speed is set by the stop screw on the throttle lever.

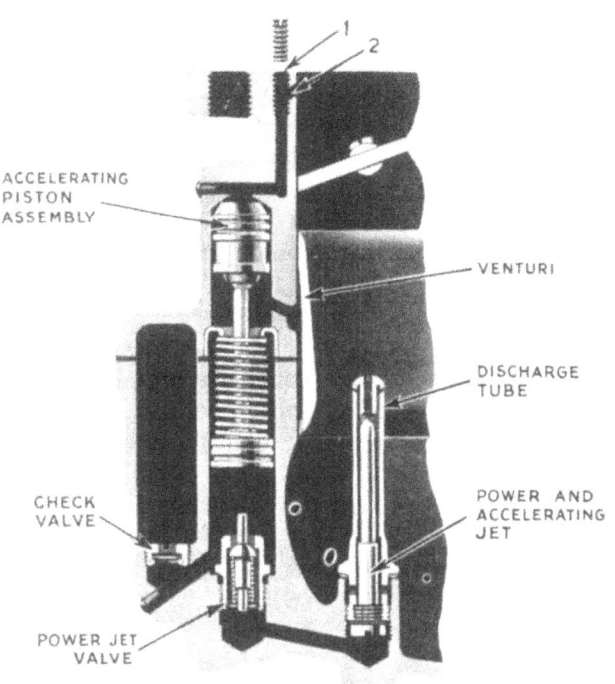

Figure 2

FULL POWER AND ACCELERATION

Full power, either for top speed or hard pulling, requires a richer mixture than part throttle operation. So does acceleration.

This additional richness of mixture is provided by combined accelerating and economizing systems operated by the vacuum above the throttle valve.

There is a plunger pump to force fuel into the air stream; a check-valve to prevent fuel from being forced back into the fuel bowl; and an economizer valve to control the additional fuel flow. The suction above the throttle holds the pump at the top of the pump well when the throttle is partially closed. As the throttle is opened the suction decreases, releasing the pump which drops to the bottom of the well, forcing fuel ahead of it.

The economizer valve is opened as the pump nears the bottom of the well. This opens a passage for the accelerating charge and, if the throttle is held open, for the additional ration of fuel necessary for full speed or power.

ECONOMY

As the throttle is closed the pump is lifted by the increased suction, so the fuel flow is reduced for anything less than full-load operation.

This vacuum type accelerating and economizing system may be used to advantage with a governor. In this case the carburetor throttle valve is usually wide open, the speed being controlled by the governor valve. By "bridging" the governor with a suction line the pump is actuated by the suction above the controlling governor valve and economizer action is thus retained.

STARTING

The Idling System acts as a priming device because when the engine is at rest the idling jet is submerged in the fuel that fills the well. The throttle should be slightly opened as this results in a very strong suction on the idling jet. The fuel passing at high velocity over the edge of the throttle plate is finely atomized and the high vacuum instantly vaporizes and mixes it with the air. This will ensure the first few explosions. With the usual manually controlled strangler, it is sometimes difficult to keep the engine running. To overcome this, Zenith uses a Spring-Loaded Strangler.

The strangler shaft is "off-center" so that engine suction tends to pull it open. A spring tends to pull the strangler shut, but except at cranking speeds, the spring is the weaker of the two forces. Accordingly, as the engine is speeded up or slows down, the strangler opens and closes, always being in a position to deliver just the right amount of air.

This prevents over-choking and crankcase dilution and ensures continued running even in the coldest weather.

The strangler control is pulled out as usual for starting. It is left alone or pushed in slightly until the engine warms up, then pushed into the open position. No "jiggling" of the control is necessary.

FLOAT LEVEL

The float level is set at the Factory and should not be changed. If leaking should occur, clean the float valve and if necessary, replace the valve and seat or the float if the arm is worn.

SPECIFICATIONS - FOR HERCULES JXD ENGINE

```
Model 457-2
Part No. 6UG450            Idle Jet            15
Venturi  34                Power and Acc Jet   18
Main Jet 35 (Adjustable)   Cap Jet             34
Compensator 32             Fuel Valve Seat     54
```

PARTS LIST

No. 869

THE AUTOCAR COMPANY, ARDMORE, PENNSYLVANIA
SERVICE DEPARTMENT

BULLETIN REF. 691

ASSEMBLY 6UG450 - MODEL 457-2 ZENITH CARBURETOR

Req.	Description	Zenith Number
1	Main Jet Adj. Assembly	C71-3
1	Idle Adj. Needle	C46-6
1	Air Shutter Bracket (R.H.) (Assembly)	C109-2
1	Float Bracket	CR88-2
1	Throttle Shaft Bushing	CR9-5
1	Air Sh. Bracket Clamp (Wire)	CR110-1
1	Power & Accel. Jet Fibre Washer	T56-48
1	Cap Jet Fibre Washer	T56-24
1	Channel Screw Fibre Washer	T56-5
1	Compensator Jet Fibre Washer	T56-24
1	Discharge Tube Fibre Washer	T56-2
1	Filter Plug Fibre Washer	T56-15
1	Fuel Valve Fibre Washer	T56-23
1	Main Jet Fibre Washer	T56-24
1	Main Jet Adj. Fibre Washer	T56-13
1	Union Body Fibre Washer	T56-36
1	Filter Plug	C149-22
1	Filter Screen	C150-1
1	Float Assembly	C85-6
1	Float Axle	C120-6
1	Fuel Bowl Assembly	B3-18E
1	Bowl to Body Gasket	C142-28
1	Power & Accel. Jet #18	C51-3
1	Cap Jet #34	C57-1 x 1
1	Compensator Jet #32	C52-3
1	Idling Jet #15	C54-1
1	Main Jet #35	C52-4
1	Air Shutter Lever (Assembly) R.H.	C106-2
1	Throttle Clamp Lever	C24-10 x 8
1	Throttle Stop Lever L.H.	CR28-33 x 5
1	Bowl to Body Screw Lock Washer	T43-103
1	Venturi Set Screw Lock Washer	T41-10
1	Air Sh. Shaft Nut Lockwasher	T45-8
1	Retainer Screw Lockwasher	T43-6
1	Clamp Screw Nut	T21S8
1	Air Shutter Shaft Nut	T22S8
1	Taper Pin (Thrust Washer)	CT63-2
1	Throttle Stop Pin	CR121-10
1	Air Shutter Plate (Assembly)	C101-19
1	1/8" Pipe Plug (Gov. by-pass)	CT91-1
1	1/8" Pipe (Bowl Drain)	CT91-1
1	Pump & Vacuum Piston Assembly	C36-12
1	Packing Ring R.H.	C130-16
1	Air Shutter Bracket Clamp Screw	T1S8-10
1	Swivel Screw	T1S8-6

Parts List No. 869
Bulletin Ref. #691

Req.	Description	Zenith Number
2	Bowl to Body Screw (Assembly)	T8S31-16
1	Channel Screw	C138-61
2	Bracket Screw (Assembly)	C140-2
1	Throttle Lever Clamp Screw	T8S10-9
2	Throttle Plate Screw	C136-12
1	Throttle Stop Screw	T8S10-15
1	Venturi Set Screw	T1S10-6
2	Air Sh. Retainer Screw	T11S6-4
1	Air Shutter Shaft	C105-130
1	Throttle Shaft & Lever Assembly	C29-245
1	Idle Adj. Needle Spring	C111-17
1	Throttle Stop Screw Spring	C111-62
1	Air Shutter Lever Swivel	CR134-1
1	Throttle Body Assembly	B2-52A-1
1	Air Shutter Shaft Thrust Washer L.H.	C130-4
1	Discharge Tube (Assembly)	C66-5
1	Union Body	C148-9A
1	Check Valve (Assembly)	C41-9
1	Fuel Valve #55	C81-3
1	Power Jet Valve (Blank) (Assembly)	C97-10
1	Main Venturi #34	C38-17
1	Swivel Washer	CT52-1
1	Packing (Throttle Shaft) Washer	CT57-2
1	Progressive Well	C76-21

SERVICE BULLETIN No. 594

THE AUTOCAR COMPANY, ARDMORE, PENNSYLVANIA
SERVICE DEPARTMENT

SUBJECT: FUEL PUMP ASSEMBLY (AC TYPE B #1521117) HERCULES JX ENGINES

OPERATION

Rotation of camshaft and eccentric actuates rocker arm D which is pivoted at E and which pulls the pull rod F, together with diaphragm A held between metal discs B downward against spring pressure C thus creating a vacuum in pump chamber M. Fuel from the rear tank will enter at J into sediment bowl K and through strainer L and suction valve N into pump chamber M. On the return stroke, spring pressure C pushes diaphragm A upward forcing fuel from chamber M through pressure valve O and opening P into the carburetor.

When the carburetor bowl is filled the float in the float chamber will shut off the inlet needle valve, thus creating a pressure in pump chamber M. This pressure will hold diaphragm A downward against the spring pressure C and it will remain in this position until the carburetor requires further fuel and the needle valve opens. Spring S is merely for the purpose of keeping rocker arm D in constant contact with eccentric H to eliminate noise.

SERVICE HINTS

Possible troubles, method of locating them and remedies are listed below. In some instances trouble is attributed to the fuel pump which in reality is caused by some other condition such as lack of gasoline in tank, leaky, broken or kinked tubing or loose connections on lines. These points should be carefully checked to avoid the needless replacement of fuel pumps.

If the pump fails to deliver fuel to carburetor, first, check to insure that glass sediment bowl K is tight and that the cork gasket under the bowl is flat in its seat and not broken, second, inspect and clean screen L and make certain that the cork gasket is properly seated when reassembling, third, tighten valve plugs securely replacing valve plug gaskets if necessary and fourth, remove valve plugs, take out check valves N and O and wash in gasoline. If valves are damaged or warped replace them. Examine valve seats to make certain that there are no irregularities to prevent proper seating of the valves and then reassemble valves and plugs after making certain that the valve springs are properly attached to the valve plugs. If fuel leaks through vent hole in body, the diaphragm A is either broken or the fuel is leaking through the diaphragm gaskets. Check the diaphragm nut and gasket and if necessary, replace diaphragm. If carburetor floods, adjust or repair carburetor needle valve seat but do not alter fuel pump. For more extensive repairs the fuel pump unit should be referred to the nearest AUTOCAR or AC representative for repair or exchange.

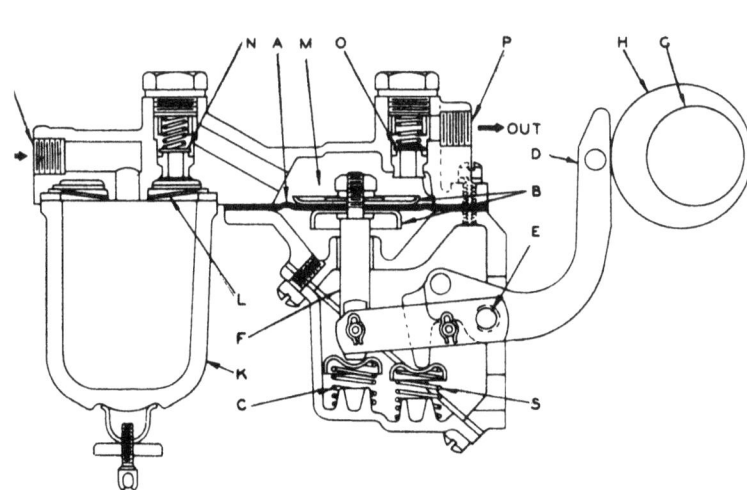

PARTS LIST

No. 862

THE AUTOCAR COMPANY, ARDMORE, PENNSYLVANIA
SERVICE DEPARTMENT

BULLETIN REF. 594

ASSEMBLY 6BL4510 - AC TYPE "B" FUEL PUMP ASSEMBLY - 1521117

Req.	Description	AC Spark Plug No.
2	Valve Plug	855135
2	Valve Plug Gasket	855136
2	Valve	855003
2	Valve Spring	856270
1	Top Cover & Valve Seat Assembly	1523358
1	Glass Bowl	854004
1	Bowl Gasket	854003
1	Screen	854009
1	Bowl Seat	854005
1	Bail & Screw	854016
1	Bail Thumb Nut	855763
1	Diaphragm (4 pieces)	855035
1	Upper Diaphragm Protector Washer	1521194
1	Lower Diaphragm Protector Washer	855078
1	Pull Rod	855250
1	Pull Rod Nut	855213
1	Pull Rod Lock Washer	855390
1	Diaphragm Alignment Washer	855029
1	Pull Rod Gasket	855012
1	Body	1523352
1	Rocker Arm	855777
1	Rocker Arm Pin	1521289
1	Rocker Arm Pin Washer	1521288
1	Bottom Cover	855228
1	Bottom Cover Gasket	855229
3	Bottom Cover Screw	132108
2	Rocker Arm & Diaphragm Spring	855253
2	Spring Cap	855532
2	Link Pin	855016
4	Link Pin Clip	855017
2	Link	855374
6	Top Cover Screw	855493
6	Top Cover Screw Lock Washer	855064

SERVICE BULLETIN No. 688

THE AUTOCAR COMPANY, ARDMORE, PENNSYLVANIA
SERVICE DEPARTMENT

SUBJECT UNITED AIR CLEANER 6ZBM0330 FOR HERCULES JX ENGINE

This air cleaner, maker's no. H80-9471, is an oil bath type with the intake around the upper edge and the discharge at the bottom center.

CAUTION

To service cleaner remove only the wing nut. Never use tools to tighten wing nut, use fingers only.

Remove wing nut and lift the upper section of the cleaner out of the sump. Clean and refill sump with the same grade of oil as is being used in the engine, filling the sump to the mark stamped in the wall of the sump, marked "OIL LEVEL".

This operation should be completed each time the oil in the engine crankcase is changed or more frequently under severe dust conditions.

Occasionally, wash the upper section in a pail of gasoline.

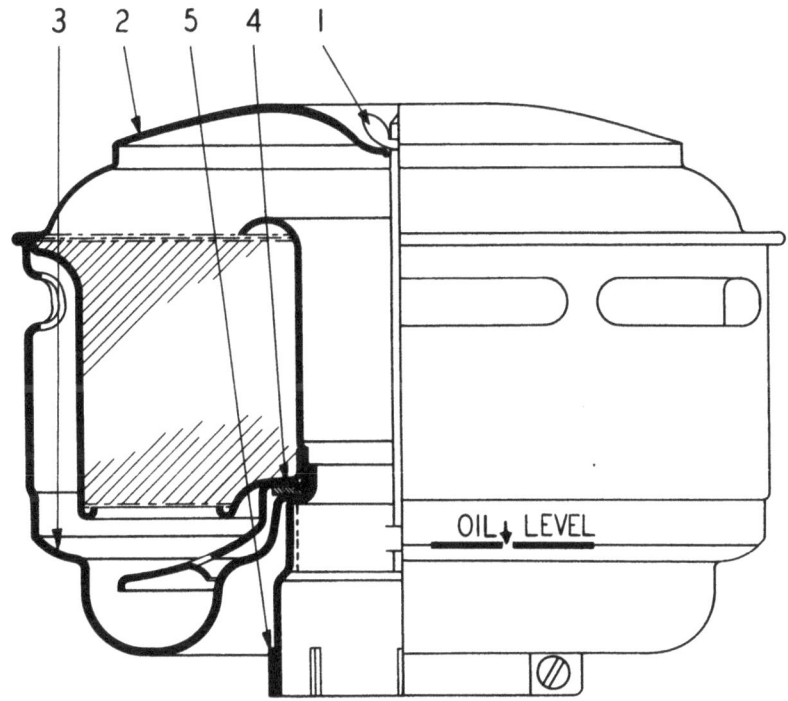

Sketch Ref. No.	Req.	Description	United Co. Number
	1	Air Cleaner Assembly - Complete	H80-9471
1	1	Wing Nut	8438
2	1	Upper Half Assembly	8549
3	1	Lower Half Assembly	9506
4	1	Gasket	8445
5	1	Clamp Assembly	7274

PARTS LIST

No. 852

THE AUTOCAR COMPANY, ARDMORE, PENNSYLVANIA
SERVICE DEPARTMENT

BULLETIN REF.

ASSEMBLY — EXHAUST SYSTEM - MODEL U-2044 - U.S.A. 1940

Part No.	Req.	Description
7TE310	1	Muffler Assembly
7SA329	2	Muffler Clip
7UK403	1	Exhaust Pipe
7BL004	1	Exhaust Pipe Flange - Hercules #40314A
7BL022	1	Exhaust Pipe Flange Gasket - Hercules #40028A
7T144	2	Brass Nut - 1/2"-13
7SA242	4	Lock Washer - Sheet Metal
S-1871	2	1/2"-13 - Bolt - 2-1/2" Long
12UK256	1	Muffler Bracket - Front
12DU256	1	Muffler Bracket - Rear
7UG309	1	Muffler Tail Pipe
7UG320	1	Flame Arrestor
7UG329	1	Flame Arrestor Clamp
12UG356	1	Flame Arrestor Bracket

SERVICE BULLETIN No. 702

THE AUTOCAR COMPANY, ARDMORE, PENNSYLVANIA
SERVICE DEPARTMENT

SUBJECT IGNITION SYSTEM - DISTRIBUTOR TYPE DR-1110513 FOR HERCULES JX ENGINE

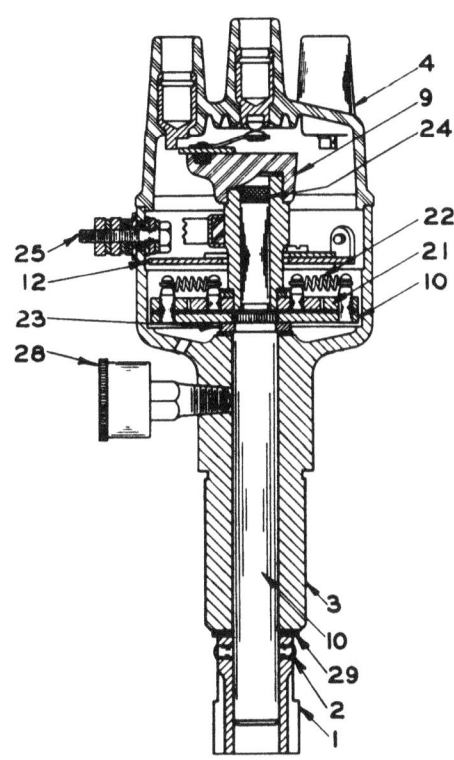
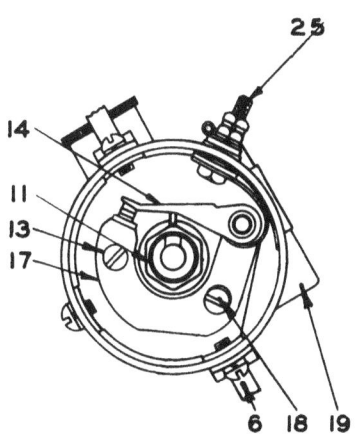

DISTRIBUTOR

The type 1110513 six-cylinder distributor is equipped with a six lobe cam and single breaker arm. Spark advance is semi-automatic. The manual control provides a maximum of 15° advance. The automatic mechanism, which operates entirely independent of the manual advance, cuts in at 600 R.P.M. (engine speed) and provides a gradual advance up to 14° advance at maximum governed speed of the engine. This distributor should be set to break with the engine flywheel set at top dead center and full manual advance.

BREAKER POINTS

The contact points should be set and maintained between .018 and .024 clearance. Points set too close will burn and pit rapidly while points with too much separation will cause missing at high speed. To adjust points, loosen breaker hold-down screw 13 and turn slotted eccentric screw 18 until proper setting is obtained, tighten hold-down screw and recheck setting.

The contact points should be inspected for wear and adjustment every 5000 miles. Contacts should be kept smooth, parallel and clean. To dress contacts it is essential to remove them from the distributor. Use an oil stone in preference to a file for this purpose. In dressing points it is only necessary to remove high spots and brighten the surface. It is not necessary to remove all traces of pit marks which in some cases will waste tungsten metal from which the points are made.

The spring tension of the contact arm should measure 17 to 21 ounces.

CONDENSER

A cartridge type condenser 19 is mounted outside the distributor housing. The condenser is placed across the points to reduce arcing and prevent excessive burning. It also assists the ignition coil to deliver a snappy spark. The condenser requires no service attention other than to see that the lead connection is clean and that it has a good ground connection.

LUBRICATION

The distributor shaft is lubricated by means of the grease cup 28 located on the side of the housing. This cup should be turned down one turn approximately every 1000 miles.

A small amount of vaseline should be applied to the face of the cam every 5000 miles to prevent excessive wear on the contact arm rubbing block.

Remove rotor 9 and apply a few drops of oil to the felt wick inside the cam every 5000 miles.

IGNITION TIMING

Use #1 cylinder for this operation and refer to the timing opening which is a 7/8" hole located on the left hand side of the bell housing. Also remove distributor cap and rotor. Set the manual spark control in the fully advanced position. (Be sure that excessive lost motion is eliminated from the mechanism in order to insure full advance on the distributor).

Crank engine until #1 cylinder reaches the compression stroke and then, while watching the timing opening in the bell housing, continue to the D.C. mark on the flywheel. With the D.C. mark in position directly opposite the opening, the contact points should be set to open with the rotor in line with the contact in the distributor head leading to the #1 cylinder.

To check contact point opening, connect a test light (6-volt 3 C.P.) between the low tension post 25 and ground. With the ignition switch on, the test bulb will light up as the points are opened, at which time the flywheel D.C. mark should be directly opposite the opening.

If the flywheel marking is ahead or behind the center of the opening, reset the distributor.

Firing order of cylinders is 1-5-3-6-2-4

IGNITION COIL

The ignition coil is for the purpose of converting the low voltage current from the storage battery into high voltage current that will jump the gap in the spark plugs. It consists essentially of two circuits, namely, primary and secondary.

TESTING COIL

First: To determine whether the current is reaching the coil. With the circuit breaker contacts separated, turn on the ignition switch and hold one end of a wire against B terminal of the coil. Brush the other end against a metal part of the engine. Failure to obtain a flash indicates loose connection or open circuit in the wiring or a defective switch.

Second: To determine whether the current is passing through the coil - Primary Circuit -- With the circuit breaker contacts in the distributor open, turn on the ignition switch. Then hold one end of a wire against the other terminal of the coil and brush the other end against a metal part of the engine. If no flash occurs, the coil is open circuited and should be replaced.

Third: If the current passed through the coil on the second test, see whether the coil will give a spark - Secondary Circuit -- Crank the engine until the circuit breaker contacts close. Hold a metal object against the base of the coil and within 1/8" of the high tension terminal on the top of the coil. Now separate the circuit breaker contacts by moving both breaker arms quickly with the finger. If no spark is obtained from the high tension terminal, the coil is damaged and should be replaced.

SPARK PLUG - Spark plugs should be occasionally taken apart, cleaned and adjusted. Misfiring is frequently caused by dirty or damaged plugs. Gaps should be adjusted to .022".

PARTS LIST

No. 867

THE AUTOCAR COMPANY, ARDMORE, PENNSYLVANIA
SERVICE DEPARTMENT

BULLETIN REF. 702

ASSEMBLY 8UG0470 - DISTRIBUTOR TYPE DR-1110513 FOR HERCULES JX ENGINE

Sketch Ref. Number	Part No.	Req.	Description	Delco-Remy No.
	8UG0470	1	Distributor & Gear Assembly	1110513
1	2BLA0855	1	Distributor Gear	
	8SA0324	1	Distributor Gear Spacer Washer	811912
2		1	Distributor Gear Pin	1857492
	8SA2120A	1	Ignition Coil	528C
3	8BLA0260	1	Distributor Housing	820668
4	8BLA069	1	Distributor Cap	1838100
6	8BLA028	2	Distributor Cap Spring	816801
	8BLA0329	2	Distributor Cap Spring Support	826476
	8SCM0302A	3	Breaker Plate and Cap Spring Support Screw	115607
9	8BLA035	1	Rotor	816774
10		1	Mainshaft & Weight Plate	1836591
11	8BLA0210	1	Cam	820989
12	8BLA0317	1	Breaker Plate	826474
13	8SA0318	1	Contact Support Fastening Screw	816784
14	8SA057	1	Breaker Lever & Point	813238
17	8BLA009	1	Contact Point & Support	1847341
19	8BLA0288	1	Condenser	1869706
	8BLA0302	1	Condenser Attaching Screw	115417
21	8BLA0230	2	Weight	821596
22		2	Weight Spring	1882978
	8SA0322	2	Weight Washer	811124
23	8BLA0324	1	Spacer Washer (Under Weight Plate)	810085
24	8BLA0325	1	Felt Washer	816803
25	8BLA0472	1	Terminal Screw	820987
	8BLA0526	2	Terminal Screw Nut	120622
	8BLA0524	1	Terminal Screw Bushing Washer	821033
	8BLA0464	1	Terminal Screw Washer	807716
	8BLA0465	1	Terminal Screw Insulation Washer	816806
	8BLA0466	1	Terminal Screw Insulation Washer	820988
	8BLA0327	1	Terminal Screw Insulation Strip	820991
28	8SA0287	1	Grease Cup	805579
29	8SA0294	As	Shim Washer .005 Thick	811074
29	8SA0295	As	Shim Washer .010 Thick	810078
	8BL2311	1	Advance Control Arm Assembly	
	8BL0284	1	Advance Control Arm	1GB-1171
		1	Advance Control Screw	1G-750
		1	Advance Control Screw Spring Washer	1G-687
		1	Advance Control Screw Washer-Top	1G-688
		1	Advance Control Screw Washer-Bottom	1G-688A
		1	Advance Control Screw Nut	8X-163

Parts List No. 867
Bulletin Ref. #702

Sketch Ref. Number	Part No.	Req.	Description	Delco-Remy No.
		1	Advance Control Arm Clamp Screw	8X-707
		1	Advance Control Arm Clamp Screw Nut	8X-146
		1	Advance Control Arm Screw	8X-870
	8BN0610	1	Ignition Cables Assembled	
	8TE018	6	Ignition Wire Hook	
	8D023	6	Spark Plug	Champion #H-10
	8TE0449	As	Ignition Wire	

SERVICE BULLETIN   No. 696

THE AUTOCAR COMPANY, ARDMORE, PENNSYLVANIA
SERVICE DEPARTMENT

SUBJECT   TIMKEN MODEL F-551-TW FRONT DRIVING AXLE

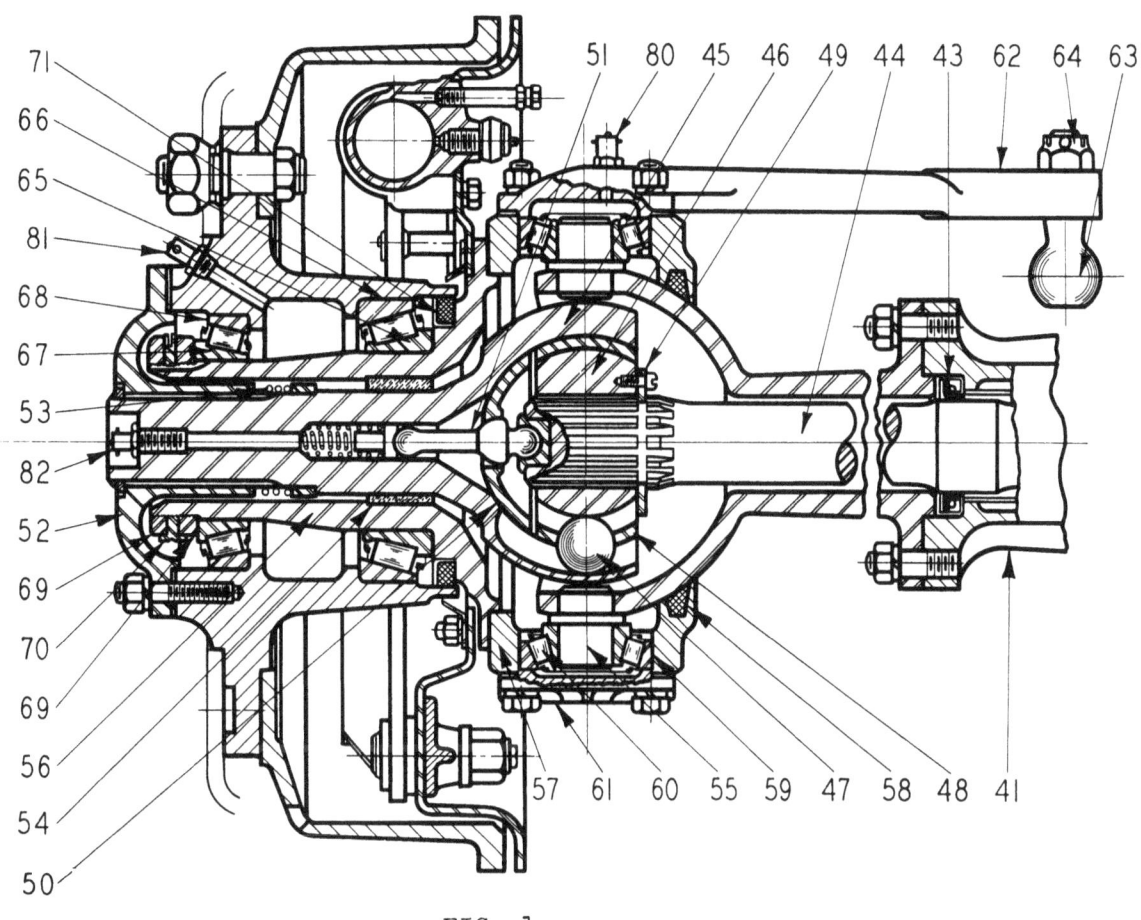

FIG. 1

GENERAL

These axles are equipped with a double reduction gear mechanism of conventional type mounted in a gear carrier and detachable as a unit from the axle housing. The gear mechanism is illustrated in Figure 2.

The axle housing is jointed at the outer ends to permit steering action at the wheels necessitating the use of universal joints on the driving axles inside the steering knuckles. The steering knuckle, universal joint and front wheel hub details are illustrated in Fig. 1.

LUBRICATION

Gear Carrier - The lubrication system in the gear carrier provides a positive oil feed to the pinion and jackshaft bearings. The differential unit runs directly in the lubricant reservoir. For proper lubrication of this unit use a heavy oil conforming to the following specifications. DO NOT USE GREASE.

For summer, use SAE No. 140 oil - for winter, SAE No. 90 Oil

Check level of lubricant weekly or every 1000 miles and add oil as required. Fill to level plug in front of housing.

Every three months or every 10,000 miles, drain axle case, wash out with kerosene and refill to level plug.

Bulletin #696

LUBRICATION - continued

Universal Joints (Driving Axles) - Lubricate joints through fittings 82 (Fig. 1) in each hub weekly or every 1,000 miles of service using SAE No. 140 Oil for summer, SAE No. 90 oil for winter. Fill up to level plug in housing.

Wheel Bearings - Lubricate wheel bearings through fittings at 81 (Fig. 1) in each wheel monthly or every 5,000 miles of service. Use U.S. Navy Grade No. 2 grease.

Steering Knuckle - Cross Tube Ends - Lubricate knuckle and cross tube ends (four points) through fittings weekly or every 1,000 miles of service. Use heavy SAE No. 140 oil.

FRONT WHEEL ALIGNMENT

Toe-in - The front wheels should be set at zero to 1/8" wide at front measured at hub height and at corresponding points on the inside of the tires. To adjust toe-in, remove clevis pin and loosen clamp bolts in right hand clevis so that clevis can be threaded in or out on cross tube.

Caster Angle - A comparative high angle is necessary to compensate for the torque reaction of driving mechanism. The axle is assembled with a positive caster angle of 4° to 5-1/2°.

Camber Angle - The wheel camber is zero and cannot be altered through any adjustment.

Steering Angle - Stop screws located on the spindle castings adjacent to the cross tube ends control the steering angle. The stop screws should be adjusted at each spindle to provide a maximum turning angle in both directions of 30°.

ADJUSTMENTS

Front Wheel Bearings
See Fig. 1
The wheels are mounted on Timken tapered roller bearings. These bearings are adjusted by means of conventional double nut device 69 and 70.

In making an adjustment turn the adjusting nut 69 tight against the outer bearing until the wheel binds, then back off nut one half turn or until wheel rotates freely. Apply lock ring 70 and tighten jam nut 69. Be sure that wheel rotates

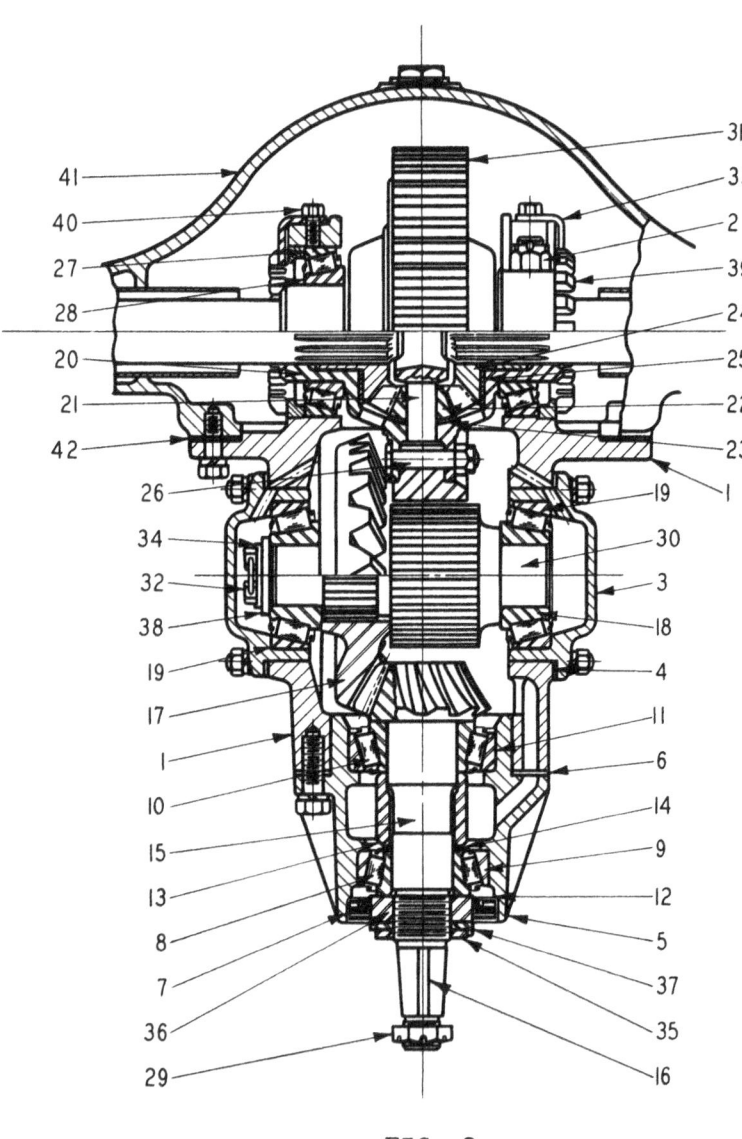

FIG. 2

ADJUSTMENTS - Front Wheel Bearings - continued

freely without excessive play.

Spindle Bearing (See Fig. 1)

The spindle is mounted on Timken tapered roller bearings 60, held in position by caps 61 at top and bottom, and adjusted by means of shims under bearing caps. In making adjustments, remove or add an equal number of shims at top and bottom of spindle so that the centralized position of the spindle and universal joint is not disturbed.

Gear Carrier (Fig. 2)

The pinion bearings 8-10 which are of the tapered roller type are adjusted by means of shims 14 located between the outside bearing cone 8 and the spacer on the pinion shaft. Remove shims until bearings bind, then add a .001" or .002" shim to provide an adjustment of .000" - .002" tight.

To adjust mesh of pinion gear add or remove shims 6 under pinion cage. (For details on correct bevel gear adjustment refer to Service Bulletin #315).

Adjust jackshaft bearings 18 by removing or adding shims 4 under bearing caps at each end of the shaft. Remove shims to eliminate all play and adjust to .000" - .002" tight. Upon completion of the bearing set-up, check tooth contact of the bevel gear and pinion (Refer to Service Bulletin #315 for details on gear setting, and shift the jackshaft endways until correct tooth contact is obtained. In moving jackshaft, transpose shims 4 from one cap to the other so that bearing adjustment will not be changed.

The differential is mounted on Timken bearings 28 which are adjusted to .000" - .002" tight with adjuster ring 39, held in place with lock 33.

PARTS LIST No. 848

THE AUTOCAR COMPANY, ARDMORE, PENNSYLVANIA
SERVICE DEPARTMENT

BULLETIN REF. 696

ASSEMBLY — TIMKEN MODEL F-551-TW-X1 FRONT DRIVING AXLE

Sketch Ref. Number	Part No.	Req.	Description	Timken Number
			GEAR CARRIER GROUP	
1	4BA0170	1	Mounting Assembly (Gear Carrier Assembly) Specify Ratio	G-316
1	4UG0920	1	Mounting with Caps (Gear Carrier)	A2-3800-K-63
	4NFG0122	4	Differential Support Cap Stud	4-X-177
2	S-810	4	Differential Support Strap Stud Nut	N-210
3	4BLA0460	2	Mounting Side Cap & Cup Assembly	A3838-Y-25
4	4BLA097	As	Mounting Side Cap Shim-Medium	2803-T-150
4	4BLA096	As	Mounting Side Cap Shim-Thick	2803-V-152
4	4BLA098	As	Mounting Side Cap Shim-Thin	2803-U-151
4	4BLA0143	2	Mounting Side Cap Gasket-R.H.	2808-F-58
	4NFG01084	14	Mounting Side Cap and Cage Stud	4-X-186
	S-1420	14	Mounting Side Cap and Cage Stud Nut	N-18
	S-4157	1	Mounting Inspection Plug	P-120
5	4UG040	1	Pinion Bearing Cage & Cup Assembly	A3826-Z-26
6	4BLA098A	As	Pinion Bearing Cage Shim-Thin	2803-Q-147
6	4BLA097A	As	Pinion Bearing Cage Shim-Medium	2803-R-148
6	4BLA096A	As	Pinion Bearing Cage Shim-Thick	2803-S-149
7	4UG084	1	Pinion Bearing Cage Oil Seal	A1805-F-162
8	4UG01232	1	Pinion Bearing Cone-Front	49175
9	4RL0137	1	Pinion Bearing Cup-Front	49368
10	4DFL0138	1	Pinion Bearing Cone-Rear	59200
11	4RL0142	1	Pinion Bearing Cup-Rear	59412
12	4UG0762	1	Pinion Bearing Spinner Ring	1829-U-21
13	4UG0584	1	Pinion Shaft Bearing Spacer	1844-W-23
14	4UG096	As	Pinion Shaft Bearing Shim	2803-L-142
14	4UGA096	As	Pinion Shaft Bearing Shim	2803-M-143
14	4UH096	As	Pinion Shaft Bearing Shim	2803-N-144
14	4UHA096	As	Pinion Shaft Bearing Shim	2803-P-146
15	4DK035	1	Bevel Pinion Shaft (10-T)	3890-V-74
16	4NK0163	1	Pinion Shaft Key	16-X-28
17	4BLA0521	1	Bevel Gear (21-T)	3889-H-60
18	4RL0141	2	Bevel Gear Bearing Cone	59175
19	4RL0142	2	Bevel Gear Bearing Cup	59412
	4UGA010	1	Differential & Ring Gear Assembly (Specify Ratio)	A3-3835-Y-51
	4UG010	1	Differential Assembly	A2-3835-Y-51
20	4BMC0220	1	Differential Cases Assembly	A3835-Z-52
21	4BLA058	1	Differential Spider	3878-B-54
22	4UG055	4	Differential Bevel Side Pinion	2233-M-429
23	4BMA0751	4	Differential Bevel Side Pinion Thrust Washer	1229-C-861

Page 80

Parts List No. 848
Bulletin Ref. #696

Sketch Ref. Number	Part No.	Req.	Description	Timken Number
			GEAR CARRIER GROUP (Continued)	
24	4UG057	2	Differential Bevel Side Gear	2234-K-427
25	4BM0638	2	Differential Bevel Side Gear Thrust Washer	1229-Z-728
26	4BLA018	8	Differential Case Bolt	15-X-180
27	4D2157	2	Differential Bearing-Cup	472-A
28	19SKB0443	2	Differential Bearing-Cone	482
29	4A081	1	Bevel Pinion Shaft Nut	13399
30	4BLA044	1	Spur Pinion Shaft (8.21 Ratio)	3891-R-96
	4BMC01061	1	Spur Pinion Shaft (6.62 Ratio)	3891-Q-121
	4BMB062B	1	Spur Gear (Ring Gear) (8.21 Ratio)	3892-S-71
31	4BMC062	1	Spur Gear (Ring Gear) (6.62 Ratio)	3892-C-133
32	4BLA0648	2	Gear Shaft Bearing Washer Stud	4-X-322
33	4BMC0125	2	Differential Bearing Adjusting Nut Lock	1820-J-10
34	4BLA028	1	Gear Shaft Bearing Stud Lock	1820-K-11
35	4UG028	1	Pinion Bearing Lock	1827-G-7
36	4UG0936	1	Pinion Bearing Adjusting Nut	1827-N-14
37	4NK01204	1	Pinion Bearing Lock Washer	1829-E-5
38	4BLA01178	1	Gear Shaft Bearing Lock Washer	1829-F-6
	4NK0425	1	Carrier Plug	1850-B-28
39	4BMC0119	2	Differential Bearing Adj. Nut	2214-R-18
	S-811	8	Differential Case Bolt Nut	N-28
40	S-2065	2	Differential Bearing Adj. Nut Lock Screw	S-265-D
	S-1349	14	Spring Washer	W-18
			FRONT AXLE HOUSING GROUP	
41	9UG0830	1	Housing Assembly	A3801-X-388
	S-4157	1	Housing Oil Filler Plug	P-120
	4UG067	1	Housing Oil Drain Plug	P-220
42	4UG0483	1	Carrier to Housing Gasket	2808-E-57
	4NFG01084	4	Carrier to Housing Stud	4-X-186
	S-19	8	Carrier to Housing Screw	S-2812
	4UH022	1	Housing Tube (Short-L.H.)	3616-D-212
	4UGA022	1	Housing Tube (Long-R.H.)	3816-C-211
	4FD0648	16	Socket to Housing Stud	4-X-173
	S-1350	16	Spring Washer	W-19
	S-830	4	Carrier to Housing Stud Nut	N-48
	S-84	16	Socket to Housing Stud Nut	N-49
43	4ZDK0395	2	Driving Axle Oil Seal Assembly	A1805-B-54
	4BL01600	1	Housing Oil Breather	A1199-E-421

Parts List No. 848
Bulletin Ref. #696

Sketch Ref. Number	Part No.	Req.	Description	Timken Number
			FRONT DRIVING AXLE & WHEEL BEARING GROUP	
44	9UK01030	1	Driving Axle & Pilot Seat Assem.- R.H. (Inner) - Long	A3802-F-370
44	9UG01030	1	Driving Axle & Pilot Seat Assem.- L.H. (Inner) - Short	A3802-G-371
45	9ZDK0820	2	Universal Drive Assembly	A3897-B-236
46	9ZDK0476	2	Universal Joint Cage	1898-F-58
47	9ZDK0477	12	Universal Joint Cage Ball	1898-G-59
48	9ZDK0484	2	Universal Joint Cage Ball Race- Inner	1898-H-60
49	9ZDK0478	2	Universal Joint & Shaft Retainer	1898-J-62
	9ZDK0479	6	Universal Joint Retainer Screw	1199-A-651
	9ZDK0504	2	Universal Joint Retainer Snap Ring	1854-H-8
50	9ZDK0481	2	Universal Joint Pilot	1898-L-64
51	9ZDK0482	2	Universal Joint Pilot Pin	1898-M-65
	9AK0486	2	Universal Joint Pilot Pin Plunger	1898-N-66
	9AK0487	2	Universal Joint Pilot Pin Plunger Spring	1898-P-68
	9ZDK0483	2	Universal Joint Pilot Pin Seat	1898-A-53
	9SK0488	2	Universal Joint Buffer Spring	1898-Q-69
52	9ZDK0435	2	Driving Flange	3870-D-4
	9ZDK0436	2	Driving Flange Gasket	2808-W-75
	S-12	4	Driving Flange Puller Screw	S-266
	S-75	4	Driving Flange Puller Screw Nut	N-36
	4ZDK0237	16	Driving Flange Stud	4-X-509
53	9ZDK0511	2	Universal Joint Spring	2858-C-3
54	9ZDK0438	2	Universal Drive Bushing	1825-G-7
	9NK0473	2	Drive Shaft Retainer Gasket	2808-H-138
	9ZDK0489	2	Universal Joint Packing	5-X-346
	9ZDK0499	2	Universal Joint Spacer	1844-D-30
			STEERING KNUCKLE GROUP	
	9ZDK0860	2	Trunnion Socket Assembly (Spindle Pin Socket)	A3897-S-201
55	9ZDK0444	4	Trunnion Socket Bearing Pin (Spindle Pin)	2847-D-56
56	9ZDK0140	2	Steering Knuckle	A3811-G-7
	9ZDK0498	2	Trunnion Socket Felt Spring	1818-E-5
57	9UG0890	1	Steering Knuckle Flange Assem-R.H.	A2-3897-N-274
57	9UGA0890	1	Steering Knuckle Flange Assem-L.H.	A3-3897-N-274
58	9ZDK0447	2	Steering Knuckle Felt Washer	5-X-290
	9ZDK0448	24	Steering Knuckle Stud	4-X-317
	9SK0449	2	Tie Rod Bolt Bushing	1825-S-71

Parts List No. 848
Bulletin Ref. #696

Sketch Ref. Number	Part No.	Req.	Description	Timken Number
			STEERING KNUCKLE GROUP (Continued)	
59	4ZDK0139	4	Steering Knuckle Bearing-Cup	41286
60	4ZDK0138	4	Steering Knuckle Bearing-Cone	41125
	9ZDK0452	1	Steering Knuckle Bearing Cap-Upper	3866-J-62
	4NFG01084	8	Steering Knuckle Bearing Cap Stud	4-X-186
61	9UG0453	2	Steering Knuckle Bearing Cap-Lower	3866-X-63
	9ZDK0454	As	Steering Knuckle Bearing Cap-Shim-Thin	2803-H-138
	9ZDKA0454	As	Steering Knuckle Bearing Cap Shim-Thick	2803-J-140
62	10UG093	1	Steering Arm	3933-E-83
63	9UGA023	1	Steering Arm Ball Stud (1-1/2" Dia)	2910-C-3
64	9UG0294	1	Steering Arm Ball Stud Nut	13-X-36
			TIE ROD GROUP	
	9UG025	1	Tie Rod	3902-Z-52
	9NK022	1	Tie Rod Yoke-R.H. (16 Thds.)	3944-B-28
	9SK021	1	Tie Rod Yoke-L.H. (12 Thds.)	3944-H-8
	9SK024	2	Tie Rod Yoke Bolt	10-X-216
	4ZGEW0974	2	Tie Rod Yoke Bolt Lock Pin	1246-L-220
	9AK0475	2	Tie Rod Yoke Nut	1827-J-88
	9ZDK0475	1	Tie Rod Yoke Lock Nut	1827-K-89
			FRONT HUB & WHEEL BEARING GROUP	
65	4FD0138	2	Front Hub Bearing Cone-Inner	594
66	4FD0139	2	Front Hub Bearing Cup-Inner	592-A
67	19SKB0445	2	Front Hub Bearing Cone-Outer	498
68	19SKB0444	2	Front Hub Bearing Cup-Outer	493
69	9ZDK028	4	Front Hub Bearing Nut	1827-H-34
70	9ZDK0175	2	Front Hub Bearing Adjusting Nut Lock Washer	1829-G-85
71	9ZDK095	2	Front Hub Bearing Felt Washer	5-X-285
73	9ZDK0178	2	Front Hub Bearing Felt Washer Retainer	1805-X-24
	9BP0807	2	Front Hub & Cups Assembly	A322-A-105
	9BP0232	2	Front Brake Drum	3819-K-11
	4A3489	10	Front Wheel Budd Stud-R.H.	12247-E
	4A3491	10	Front Wheel Budd Stud-L.H.	12248-E
	9A0216	10	Front Wheel Budd Stud Nut-R.H.	37888-E
	9A0217	10	Front Wheel Budd Stud Nut-L.H.	37889-E
	4A0499	20	Front Hub Stud Nut	13332-E
	4A0503	10	Front Wheel Budd Stud Nut-Outer-R.H.	37891-E
	4A0504	10	Front Wheel Budd Stud Nut-Outer-L.H.	37892-E

Sketch Ref. Number	Part No.	Req.	Description	Timken Number
			FRONT HUB & WHEEL BEARING GROUP (Continued)	
	4A0505	10	Front Wheel Budd Stud Nut-Inner-R.H.	10708-E
	4A0506	10	Front Wheel Budd Stud Nut-Inner-L.H.	10709-E
			FRONT BRAKE GROUP	
	4HDA0100	4	Brake Shoe Assembly	A3-3822-T-306
	4UG052	4	Brake Shoe Lining	2240-J-530
	S-1020	56	Brake Shoe Lining Rivet	X-1838
	4UG089	4	Brake Shoe Bushing	1225-U-99
	4HDA01317	4	Brake Shoe Anchor Pin	1259-M-39
	S-3753	4	Brake Shoe Anchor Pin Nut	N-112
	S-1353	4	Spring Washer	W-112
	4UG01326	12	Brake Shoe Anchor Pin Bracket Rivet	R-158
	25G0103	4	Brake Shoe "C" Washer	1229-M-221
	25SD0235	4	Brake Shoe Wear Plate	2217-Z-26
	S-4280	4	Brake Shoe Wear Plate Screw	X-1678
	9UG0458	1	Brake Shaft-R.H.	2810-N-248
	9UGA0459	1	Brake Shaft-L.H.	2810-P-250
	25G0106A	2	Brake Shaft Washer	R-5706
	4UG01353	2	Brake Shaft Snap Ring	1854-B-80
	4HDA079	2	Brake Shoe Spring	2258-B-184
	25KE021	1	Air Chamber-R.H.	220127
	25KF021	1	Air Chamber-L.H.	220128
	25KH0881	2	Slack Adjuster	215091
	4UG0900	2	Brake Shaft Bracket & Bushing Asm.	A3899-J-166
	4UG0149	2	Brake Shaft Bracket Bushing-Long	1825-H-8
	4UGA0149	2	Brake Shaft Bracket Bushing-Short	1825-J-10
	S-2085	4	Brake Shaft Bracket Bolt	S-11012
	S-1311	4	Brake Shaft Bracket Bolt Nut	N-110
			FRONT BRAKE DUST SHIELD GROUP	
	4UG0120	1	Front Brake Dust Shield-R.H.	A3836-F-136
	4UG0744	1	Front Brake Dust Shield-L.H.	A1-3836-F-136
	9UG0750	2	Front Brake Dust Shield Oil Slinger	3880-H-34

Parts List No. 848
Bulletin Ref. #696

PARTS LIST

No. 872

THE AUTOCAR COMPANY, ARDMORE, PENNSYLVANIA
SERVICE DEPARTMENT

BULLETIN REF.

ASSEMBLY 16" x 3-1/2" AIR BRAKES ON TIMKEN F-551-TW-X1-FRONT AXLE

Part No.	Req.	Description	Timken Number
4HDA0100	4	Brake Shoe Assembly	A3-3822-T-306
4UG052	4	Brake Shoe Lining	2240-J-530
S-1020	56	Brake Shoe Lining Rivet	X-1838
4UG089	4	Brake Shoe Bushing	1225-U-99
4HDA01317	4	Brake Shoe Anchor Pin	1259-M-39
S-3753	4	Brake Shoe Anchor Pin Nut	N-112
S-1353	4	Spring Washer	W-112
4UG01326	12	Brake Shoe Anchor Pin Bracket Rivet	R-158
25G0103	4	Brake Shoe "C" Washer	1229-M-221
25SD0235	4	Brake Shoe Wear Plate	2217-Z-26
S-4280	4	Brake Shoe Wear Plate Screw	X-1678
9UG0458	1	Brake Shaft-R.H.	2810-N-248
9UGA0459	1	Brake Shaft-L.H.	2810-P-250
25G0106A	2	Brake Shaft Washer	R-5706
4UG01353	2	Brake Shaft Snap Ring	1854-B-80
4HDA079	2	Brake Shoe Spring	2258-B-184
25KE021	1	Air Chamber-R.H.	220127
25KF021	1	Air Chamber-L.H.	220128
25KH0881	2	Slack Adjuster	215091
4UG0900	2	Brake Shaft Bracket & Bushing Assembly	A3899-J-166
4UG0149	2	Brake Shaft Bracket Bushing-Long	1825-H-8
4UGA0149	2	Brake Shaft Bracket Bushing-Short	1825-J-10
S-2085	4	Brake Shaft Bracket Bolt	S-11012
S-1311	4	Brake Shaft Bracket Bolt Nut	N-110
4UG0120	1	Front Brake Dust Shield-R.H.	A3836-F-136
4UG0744	1	Front Brake Dust Shield-L.H.	A1-3836-F-136
9BP0232	2	Front Brake Drum	3819-K-11
9UG0750	2	Front Brake Dust Shield Oil Slinger	3880-H-34

PARTS LIST   No. 863

THE AUTOCAR COMPANY, ARDMORE, PENNSYLVANIA
SERVICE DEPARTMENT   BULLETIN REF.

ASSEMBLY: FRONT SHOCK ABSORBERS - U-2044 & U-4044 - U.S.A. 1940

Part No.	Req.	Description
28M201	2	Shock Absorber Stud (Top)
28MA201	2	Shock Absorber Stud (Bottom)
S-77	8	1/2"-13 Hex Nut
S-1349	4	1/2" Spring Washer
S-390	4	Washer (17/32" I.D. - 1-3/8" O.D. - 1/16" thick)
S-2198	4	3/32" Cotter Pin
28UG202	1	Clip Plate (R.H.)
28UG203	1	Clip Plate (L.H.)
28M205	4	Shock Absorber Stud Spacer
28MA010	2	Shock Absorber
28UG211	1	Shock Absorber Frame Bracket (R.H.)
28UG212	1	Shock Absorber Frame Bracket (L.H.)
S-1872	4	1/2"-13 Hex Cap Screw 1-1/2" Long
S-77	4	1/2"-13 Hex Nut
S-1349	8	1/2" Spring Washer

SERVICE BULLETIN No. 685

THE AUTOCAR COMPANY, ARDMORE, PENNSYLVANIA
SERVICE DEPARTMENT

SUBJECT CAM AND TWIN-LEVER ROLLER MOUNTED STEERING GEAR - MODEL T-70 - 10UK0770

ADJUSTMENTS

When making adjustments free the steering gear of all load, preferably by disconnecting the drag link from the steering arm, and loosen instrument board bracket clamp on steering gear jacket tube.

If the ball thrust bearings on the cam must be adjusted, make this adjustment 1 before making the side adjustment 2.

1. <u>Adjustment on Ball Thrust Bearings on Cam.</u> Adjust to a barely perceptible drag so that the steering wheel can be turned freely (with the thumb and forefinger lightly gripping the rim).

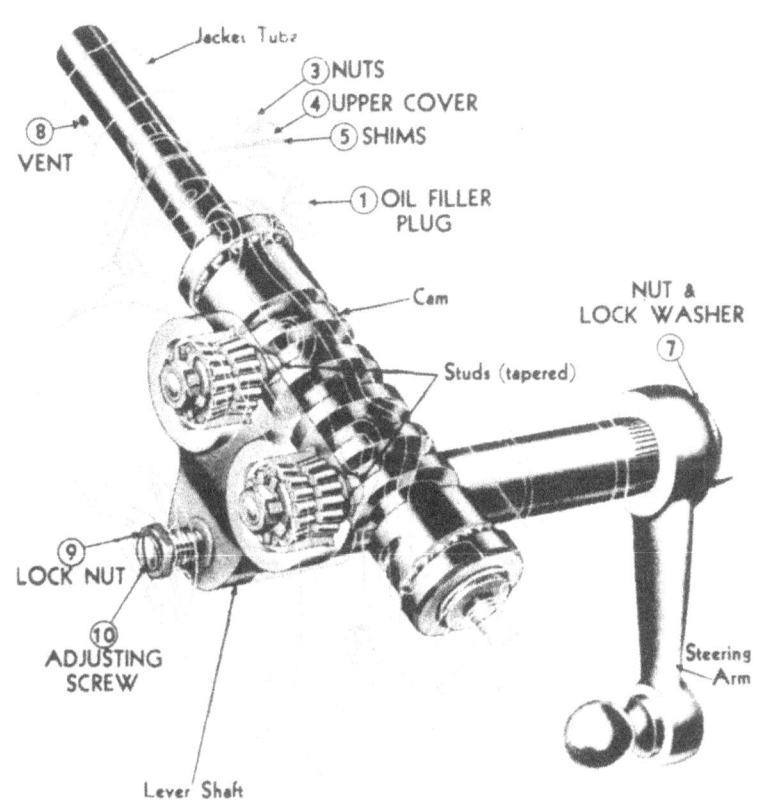

Before making this adjustment, loosen the housing side cover adjusting screw (9, 10) to free the studs in the cam groove, then unscrew the four nuts (3) and move up the housing upper cover (4) to permit removal of shims (5). (Shims are of .003" and .010" thickness.)

Clip and remove a .003 shim or more as required. Tighten all four nuts. Draw down tight. Test adjustment and if necessary, remove or replace shims until adjustment is correct.

2. <u>Adjustment for Minimum Backlash of Tapered Studs in Cam Groove.</u> Adjust so that a very slight drag is felt through the mid-position when turning the steering wheel slowly from one extreme position to the other.

Backlash of studs in the groove shows up as end play of lever shaft, also as backlash at steering wheel and at ball on steering arm.

The groove is purposely cut shallower in the straight-ahead-driving position of each stud to produce a high range in the groove (equal at each stud) that causes closer mesh of the studs in the groove through the mid-position of travel. This feature permits a close adjustment for normal straight-ahead driving and also permits take-up of backlash at this point after wear occurs without causing a bind elsewhere.

ADJUSTMENTS (Continued)

Adjust through the mid-position of stud travel. Do not adjust in positions off straight-ahead. Backlash at these turn positions is not objectionable.

Tighten side cover adjusting screw (10) until a very slight drag is felt through the mid-position. When the proper adjustment has been made, tighten the lock nut (9) and then give the gear a final test.

Secure the gear at all points loosened prior to making adjustment. Also check tightness of mounting bracket bolts and nuts, and of steering arm on lever shaft and the nut and lockwasher (7).

ADJUST STUD-ROLLER BEARING UNITS

The foregoing adjustments will suffice in nearly every instance, but in some cases it may be necessary to adjust the stud-roller bearing units in the lever shaft.

Each stud should turn with a drag with the fingers gripping the nut. New replacement bearing units should be set much tighter. The roller bearings should be pre-loaded at all times.

To Adjust: (a) Straighten out prong of locking washer. (b) Tighten nut as required, while holding stud from turning (either by spanner wrench on washer or by clamping stud). Caution: Do not nick or burr bearing surface. (c) Tap each end of stud lightly to test adjustment. (d) Lock adjustment by bending a prong of locking washer against side of nut. Bend the prong that is at right angle to a side of the nut. DO NOT USE A WASHER TWICE unless the prongs used before have been removed. (e) Wash Bearings in gasoline and make final test.

LUBRICATION

Use S.A.E. 90 oil in winter and S.A.E. 140 oil in summer.

Lubricate through pipe plug (1) every week or every 1000 miles. Fill gear until lubricant is forced out the vent hole (8) at bottom of outer tube.

GENERAL INFORMATION
COLUMN ALIGNMENT

With all supporting brackets clamped tight, turn steering wheel to see if any stiffness exists. If so, the gear is adjusted too tight or the steering column is out of alignment. THE STEERING COLUMN MUST NOT BE SPRUNG IN ANY DIRECTION. If misalignment exists, correct according to methods provided by the vehicle manufacturer.

STEERING GEAR CONNECTION WITH FRONT WHEELS

The steering gear should be in its mid-position when the front wheels are in the straight-ahead position. To check, turn the steering wheel as far to the right as possible, then rotate the wheel in the opposite direction as far as possible and note the total number of turns. Turn the wheel back just one-half of this total movement, thus placing the gear in mid-position at which point the front wheels should be in the straight-ahead position. If not, it may be necessary to remove the steering arm and shift it one spline on the shaft. Some drag links can be adjusted in length to take up minor variations.

PARTS LIST
THE AUTOCAR COMPANY, ARDMORE, PENNSYLVANIA
SERVICE DEPARTMENT

No. 827
BULLETIN REF. 685

ASSEMBLY CAM & TWIN-LEVER ROLLER MOUNTED STEERING GEAR - T-70 SERIES-10UK0770

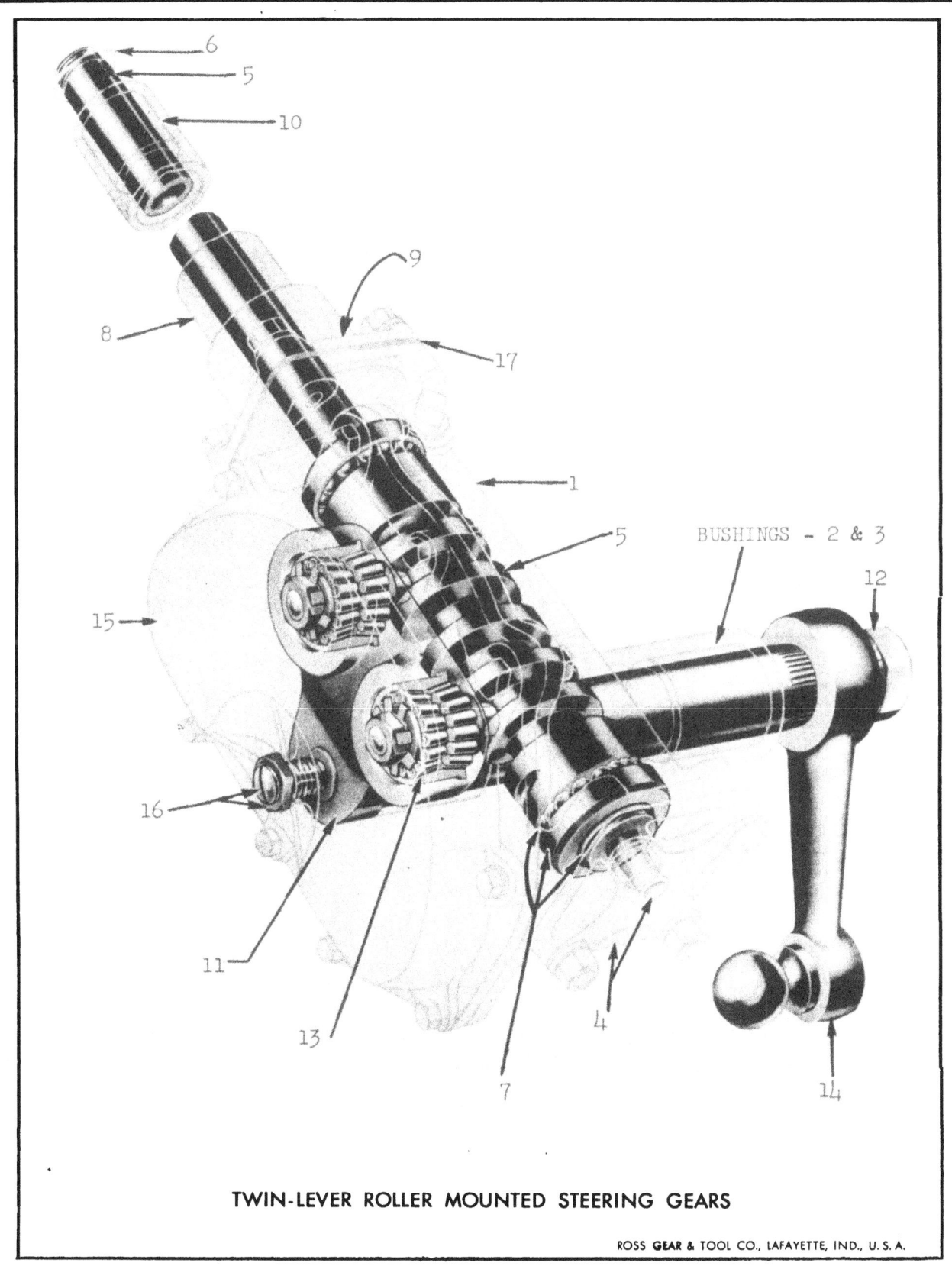

TWIN-LEVER ROLLER MOUNTED STEERING GEARS

ROSS GEAR & TOOL CO., LAFAYETTE, IND., U.S.A.

Page 89

PARTS LIST

No. 827

THE AUTOCAR COMPANY, ARDMORE, PENNSYLVANIA
SERVICE DEPARTMENT

BULLETIN REF. 685

ASSEMBLY CAM & TWIN-LEVER ROLLER MOUNTED STEERING GEAR T-70 SERIES -10UK0770

Sketch Ref. Number	Part No.	Req.	Description	Ross Gear & Tool Co. No.
	10UK0770	1	Steering Gear Assembly- Complete	T-71044
1	10UK0520	1	Housing Assembly	502955
2	10TE0215	1	Housing Bushing-Inner	069001
3	10TE0216	1	Housing Bushing-Outer	069002
4	10UK01320	1	End Cover & Oil Seal Tube Assembly	7859-13
	10UK01094	1	Housing Oil Seal Unit	032074
5	10UK0920	1	Cam and Wheel Tube Assembly with Bearings & Wheel Nut	8030-50-1/16
5	10UK0550	1	Cam & Wheel Tube Assembly with Wheel Nut	8029-50-1/16
6	10SA0251	1	Wheel Nut	C-20
	10J4210	1	Steering Wheel	
	10M2254	1	Steering Wheel Key	038601
7	10UK0224	2	Ball Cup	400021
7	10SA0227	2	Retaining Ring	400005
7	S-1670	28	Steel Balls (3/8" Dia.)	400014
8	10UK01176	1	Jacket Tube & Bearing Assembly	7789-39
9	10UK01143	1	Upper Cover	T-266000
10	10UK01097	1	Jacket Tube Bearing	065996
11	10UK0940	1	Lever Shaft & Roller Bearing Unit with Nut & Lock Washer	7920-9-3/4
11	10UK01070	1	Lever Shaft with Nut & Lock Washer	7919-9-3/4
12	10TE0314	1	Lever Shaft Nut	025003
13	10UK0930	2	Stud-Roller Bearing Unit	044987
14	10UK01370	1	Steering Arm & Ball Pin Ball Assembly	
15	10UK067	1	Side Cover	T-705000
16	10SA0593	1	Adjusting Screw	021016
16	10SA067	1	Lock Nut	025006
	10UK01178	1	Spring	401105
	10UK01253	1	Washer	028104
17	10UK01145	As	Upper Cover Brass Shim .002	033042
17	10UKA01145	As	Upper Cover Steel Shim .003	033036
17	10UKB01145	As	Upper Cover Steel Shim .010	033037
	S-1455	6	3/8"-24 Hex Head Screw x 1-1/2"	020040
	S-1445	4	3/8"-24 Hex Head Screw x 1"	T-73
	S-814	4	3/8"-24 Hex Nut	
		10	3/8"x1/8"x3/32" Lock Washer	
	S-4004	2	Pipe Plug - 3/8"-18 Std.	
	10UK0323	1	Side Cover Gasket	T-709000
	10UK0160	1	Horn Button Unit	465346
	10UK0163	1	Horn Button-Bare	450029

Parts List No. 827
Bulletin Ref. #685

Sketch Ref. Number	Part No.	Req.	Description	Ross Gear & Tool Co. No.
	10RM0166	1	Horn Button Spring	401081
	10BM0648	1	Insulating Washer	051025
	10BL0724	1	Contact Washer	029032
	10BL0725	1	Contact Spring	401082
	10RM0726	1	Contact Cap	029031
	10UK0722	1	Contact Cup	029037
	10UK01252	1	Base Plate & Retaining Segments	454895
	S-1127	3	Base Plate Screw	022020
	10BM0424	1	Cable and Terminal Assembly	7242-96

PARTS LIST

No. 865

THE AUTOCAR COMPANY, ARDMORE, PENNSYLVANIA
SERVICE DEPARTMENT

BULLETIN REF.

ASSEMBLY 10UG310 - DRAG LINK

Part No.	Req.	Description
10UG310	1	Drag Link Assembly

DRAG LINK TUBE END GROUP

Part No.	Req.	Description
10UG0162	1	Drag Link Tube
10H196	2	Drag Link Ball Seat (1-1/2" Dia. Ball)
10H195	1	Drag Link Ball Seat Spring
10H194	1	Drag Link End Plug
10UF1193A	1	Drag Link End Bolt
S-76	1	Drag Link End Bolt Nut
S-1345	2	Spring Washer

DRAG LINK ADJUSTABLE END GROUP

Part No.	Req.	Description
10ZN099	1	Drag Link Adjustable End
10H196	2	Drag Link Ball Seat (1-1/2" Dia. Ball)
10H195	2	Drag Link Ball Seat Spring
10H194	1	Drag Link End Plug
10UF1193A	1	Drag Link End Bolt
S-76	1	Drag Link End Bolt Nut
S-1345	2	Spring Washer
S-2022	1	Drag Link Adjustable End Lock Nut

PARTS LIST No. 845

THE AUTOCAR COMPANY, ARDMORE, PENNSYLVANIA
SERVICE DEPARTMENT   BULLETIN REF.

ASSEMBLY   FRAME & BRACKETS - MODEL U-2044 - U.S.A. 1940

Part No.	Req.	Description
12UK710A	1	Frame with Front & Rear Cross Members Only
12UK062	1	Frame Rail - R.H. Drilled
12UK063	1	Frame Rail - L.H. Drilled
12UK402	1	Bumper
12UK303	1	Bumper Bracket - R.H.
12UK304	1	Bumper Bracket - L.H.
12UK382	1	Front Cross Member
12UG2131	2	Front Cross Member Gusset - Top
12UG3122	2	Front Cross Member Gusset - Bottom
12UG434	1	Front Spring Anchor Bracket - Front - R.H.
12UK434	1	Front Spring Anchor Bracket - Front - L.H.
5UK0104	2	Radiator Support Bracket
12L2212	2	Front Spring Rubber Bumper
16UG5414	1	Front Cab Bracket - R.H.
10UG4177	1	Front Cab & Steering Gear Bracket - L.H.
28UG211	1	Front Shock Absorber Bracket - R.H.
28UG212	1	Front Shock Absorber Bracket - L.H.
10UG5138	1	Pedal Levers Bracket
12NJ434	2	Front Spring Shackle Bracket
12UG486	1	Rear Engine Support Bracket - R.H.
12UG3399	1	Rear Engine Support Extension Bracket - R.H.
12UG2402	1	Rear Engine Support Insulator R.H.
12UK486	1	Rear Engine Support Bracket - L.H.
12UK3399	1	Rear Engine Support Extension Bracket - L.H.
12UK2402	1	Rear Engine Support Insulator - L.H.
16UG4415	1	Rear Cab Support Cross Member
25GL303	2	Air Tank Bracket
12UK256	1	Muffler Bracket - Front
12DU256	1	Muffler Bracket - Rear
12UG356	1	Flame Arrestor Bracket
12UG4360	1	Transfer Case Front Support Assembly
12HF3164	2	Transfer Case Front Support Shim
12UG5201	1	Transfer Case Rear Cross Member Assembly
12UG4480	1	Brake Cross Shaft Cross Member
12DK445A	2	Rear Spring Front Bracket
12A125	4	Rear Spring Front Bracket Bushing
13DK3154	4	Auxiliary Spring Bracket
28UK211	2	Rear Shock Absorber Bracket - R.H.
28UK212	2	Rear Shock Absorber Bracket - L.H.
12DK454	2	Rear Spring Shackle Bracket
12UL444	1	Rear Cross Member
12UL342	4	Rear Cross Member Gusset
12CH3128	1	Towing Hook - R.H.
12CH3129	1	Towing Hook - L.H.

SERVICE BULLETIN No. 681

THE AUTOCAR COMPANY, ARDMORE, PENNSYLVANIA
SERVICE DEPARTMENT

SUBJECT   3" FRONT SPRING MOUNTING - MODELS UG & UK

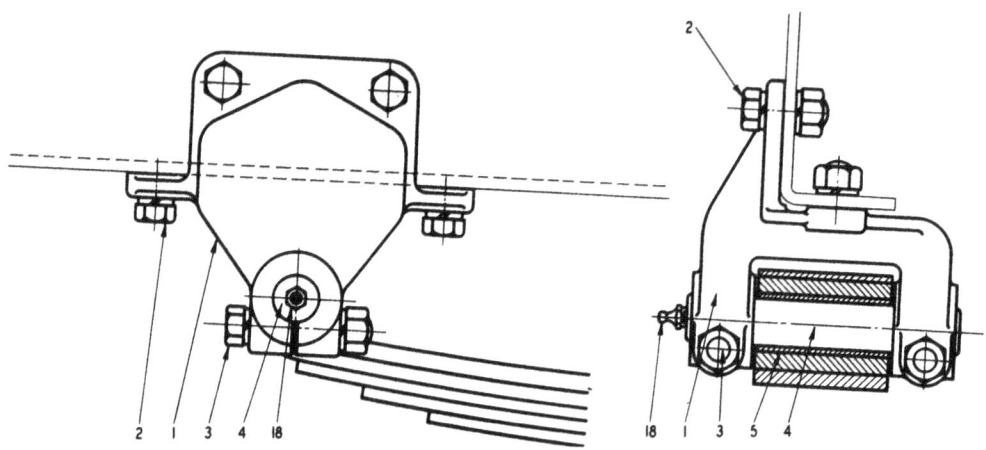

CONSTRUCTION

These models are equipped with front springs of the conventional semi-elliptic type, using a solid mounting at the front end and a shackle assembly at the rear end of the spring.

The spring eyes, brackets and shackles are all provided with renewable bushings to protect these expensive parts from wear.

MAINTENANCE

The spring clips fastening the spring to the axle should be retightened after the first 1000 miles of service to take up initial set. Periodic inspection is also recommended, as the clips must be kept tight in order to prevent leaf breakage, hold the axle in proper alignment, and to provide easy steering.

LUBRICATION

The spring pins at the front ends of the springs and the shackle pins at the rear ends should be lubricated weekly or every 1000 miles through high pressure fittings 19-20. Use a heavy oil or semi-fluid chassis lubricant.

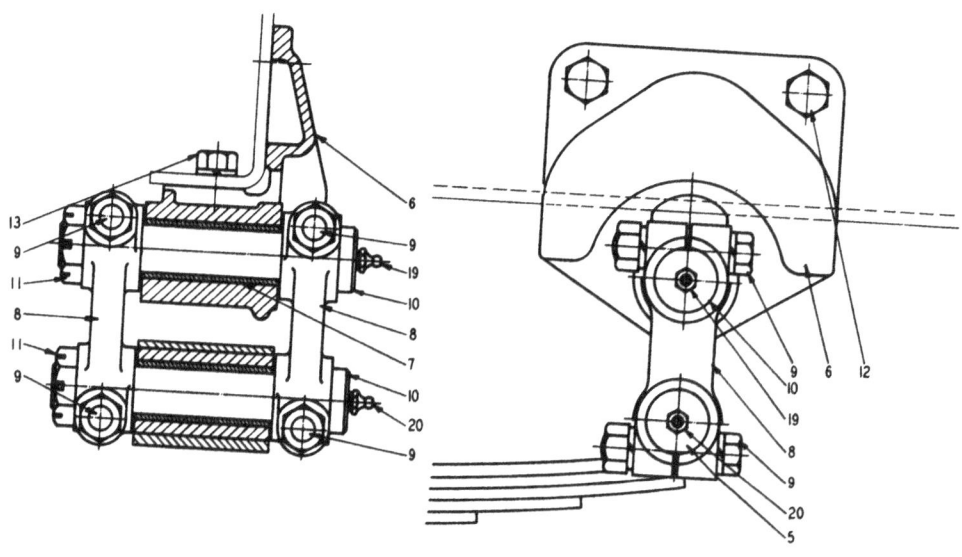

PARTS LIST No. 829

THE AUTOCAR COMPANY, ARDMORE, PENNSYLVANIA
SERVICE DEPARTMENT

BULLETIN REF. 681

ASSEMBLY 3" FRONT SPRING MOUNTING - MODELS UG AND UK

Sketch Ref. Number	Part No.	Req.	Description
			FRONT SPRING MOUNTING - FRONT
1	12UG434	1	Front Spring Bracket - Front R.H.
1	12UK434	1	Front Spring Bracket - Front L.H.
2	S-1754	8	Front Spring Bracket Bolt
3	S-4077	4	Front Spring Bracket Clamp Bolt
	S-77	12	Front Spring Bracket Bolt Nut
	S-1349	24	Spring Washer
4	13H2126	2	Front Spring Bracket Pin - Front
5	13TE226	2	Front Spring Bushing
			FRONT SPRING MOUNTING - REAR
6	12NJ0130	2	Front Spring Hook & Bushing Assembly-Rear
6	12NJ434	2	Front Spring Hook-Rear
7	13A1101	2	Front Spring Hook Bushing
8	13DF301A	4	Front Spring Shackle - Rear
9	S-911	8	Front Spring Shackle Bolt
	S-77	8	Front Spring Shackle Bolt Nut
	S-1349	16	Spring Washer
10	13DF2126	4	Front Spring Shackle Pin - Rear
11	S-3759	4	Front Spring Shackle Pin Nut
12	S-3393	4	Front Spring Hook Bolt
	S-93	4	Front Spring Hook Bolt Nut
	S-1351	8	Spring Washer
13	S-3101	4	Front Spring Hook Bolt
	S-77	4	Front Spring Hook Bolt Nut
	S-1349	8	Spring Washer

PARTS LIST

No. 825

THE AUTOCAR COMPANY, ARDMORE, PENNSYLVANIA
SERVICE DEPARTMENT

BULLETIN REF.

ASSEMBLY 13UG409 - FRONT SPRING

Width - 3"
Number of Leaves - 15
Rate Lbs. Per Inch Deflection - 1050
Spring Thickness - 4-1/16"
Free Height - 8"

Part No.	Req.	Description	Thickness
13UG409	2	Front Spring Assembly	
13UG040	2	No. 1 and No. 2 Assembled	3/8"-5/16"
13UG021	2	#1 and Bushing Assembly	3/8"
13UG022	2	#2	5/16"
13UG023	2	#3	1/4"
13UG024	2	#4	1/4"
13UG025	2	#5	1/4"
13UG027	2	#6	1/4"
13UG028	2	#7	1/4"
13UG029	2	#8	1/4"
13UG031	2	#9	1/4"
13UG032	2	#10	1/4"
13UG033	2	#11	1/4"
13UG034	2	#12	1/4"
13UG035	2	#13	1/4"
13UG065	2	#14	1/4"
13UG067	2	#15	1/4"
13TE226	4	Front Spring Bushing	
13UG0132	4	Front Spring Rebound Clip Assembly (Small)	
13UG0135	4	Front Spring Rebound Clip Assembly (Large)	
13TE0107	8	Plow Bolt	
S-850	8	Plow Bolt Nut	
S-1348	8	Spring Washer	

SERVICE BULLETIN No. 692

THE AUTOCAR COMPANY, ARDMORE, PENNSYLVANIA
SERVICE DEPARTMENT

SUBJECT   3" REAR SPRING MOUNTING - MODELS U-2044 and U-4044

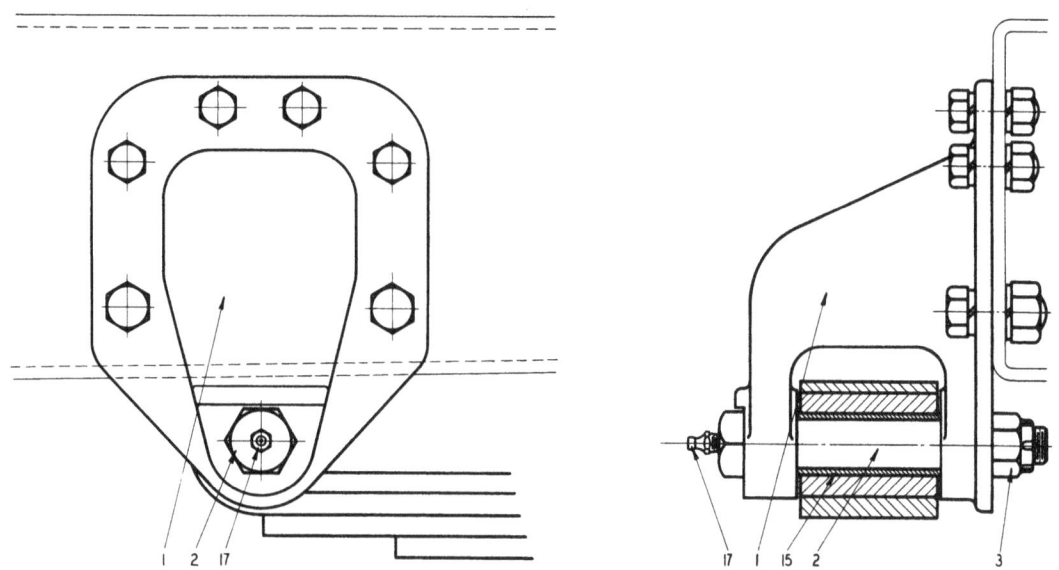

The chassis springs are semi-elliptic leaf type secured to the axle by spring clips and nuts and to the frame brackets by spring bolts and shackles.

The spring eyes, brackets and shackles are all provided with renewable bushings to protect these expensive parts from wear.

MAINTENANCE

Spring clips should be retightened after first 1,000 miles of service to take up initial set. Spring clips must be kept tight in order to prevent leaf breakage, to hold axle in proper alignment and provide easy steering.

LUBRICATION

The spring bolts and shackle pin bearings should be lubricated every week or every 1,000 miles. Use a heavy oil or a semi-fluid grease through the high pressure fittings 17-18-19. It is also good practice to apply a small quantity of water pump grease or other lubricant of heavy body to the underside of the auxiliary spring pads every week or every 1,000 miles.

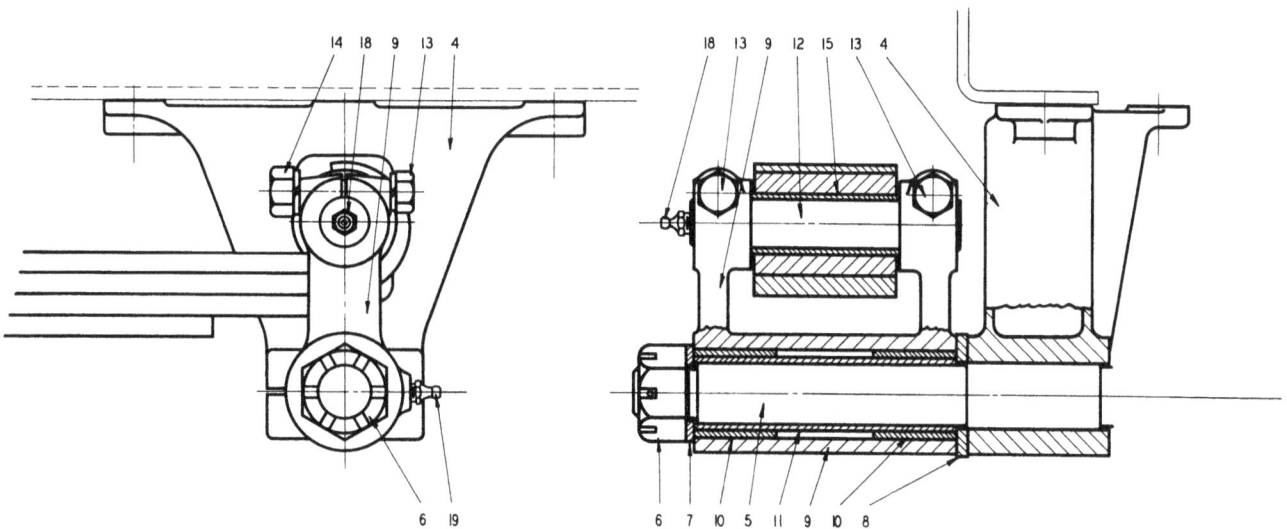

Page 97

PARTS LIST   No. 837

THE AUTOCAR COMPANY, ARDMORE, PENNSYLVANIA
SERVICE DEPARTMENT   BULLETIN REF. 692

ASSEMBLY   3" REAR SPRING MOUNTING - MODEL U-2044

Sketch Ref. Number	Part No.	Req.	Description
			REAR SPRING MOUNTING - FRONT
1	12DK445A	2	Rear Spring Bracket - Front
2	13A2115	2	Rear Spring Bolt - Front
3	S-4221	2	Rear Spring Bolt Nut
			REAR SPRING MOUNTING - REAR
4	12DK454	2	Rear Spring Shackle Bracket - Rear
5	13DK395	2	Rear Spring Cross Shaft
6	13H297	2	Rear Spring Cross Shaft Nut
7	S-979	2	Rear Spring Shackle Washer - Outer
8	13N2129	2	Rear Spring Shackle Washer - Inner
9	13H070	2	Rear Spring Shackle & Bushing Assembly-Rear
9	13H306	2	Rear Spring Shackle - Rear
10	13H299A	4	Rear Spring Shackle Bushing
11	13H298A	2	Rear Spring Shackle Spacer-Rear
12	13H2126	2	Rear Spring Pin-Rear
13	S-4077	4	Rear Spring Shackle Bolt
14	S-77	4	Rear Spring Shackle Bolt Nut
	S-1349	8	Spring Washer
15	13A1101	4	Rear Spring Bushing

PARTS LIST   No. 826

THE AUTOCAR COMPANY, ARDMORE, PENNSYLVANIA
SERVICE DEPARTMENT   BULLETIN REF.

ASSEMBLY   13UG411 - REAR SPRING

Width - 3"
Number of Leaves - 11
Rate Lbs. Per Inch Deflection - 1310
Spring Thickness - 4-1/2"
Free Height - 7-3/8"

Part No.	Req.	Description	Thickness	Length
13UG411	2	Rear Spring Assembly		
13UG050	2	1 and 2 - Assembled	7/8"	54"
13UG036	2	1 with Bushing	7/16"	54"
13UG037	2	2	7/16"	54"
13UG038	2	3	7/16"	54"
13UG039	2	4	7/16"	48-1/2"
13UG044	2	5	7/16"	43"
13UG045	2	6	7/16"	37-1/2"
13UG046	2	7	3/8"	32"
13UG047	2	8	3/8"	26-1/2"
13UG048	2	9	3/8"	21"
13UG049	2	10	3/8"	15-1/2"
13UG051	2	11	3/8"	10"
13TE226	4	Rear Spring Bushing		
13UG0127	4	Rear Spring Rebound Clip Assembly (Small)		
13UG0128	4	Rear Spring Rebound Clip Assembly (Large)		
13B0107	8	Plow Bolt		
S-850	8	Plow Bolt Nut		
S-1348	8	Spring Washer		

PARTS LIST No. 824

THE AUTOCAR COMPANY, ARDMORE, PENNSYLVANIA
SERVICE DEPARTMENT

BULLETIN REF.

ASSEMBLY 13UG4156 - REAR AUXILIARY SPRING

Width - 3"
Number of Leaves - 6
Rate Lbs. Per Inch Deflection - 2300
Spring Thickness - 2-1/4"
Free Height - 4-1/2"

Part No.	Req.	Description	Length
13UG4156	2	Rear Auxiliary Spring Assembly	
13UG0157	2	Rear Auxiliary Spring Plate No. 1	39"
13UG0158	2	Rear Auxiliary Spring Plate No. 2	36"
13UG0159	2	Rear Auxiliary Spring Plate No. 3	31"
13UG0161	2	Rear Auxiliary Spring Plate No. 4	24"
13UG0162	2	Rear Auxiliary Spring Plate No. 5	17"
13UG0167	2	Rear Auxiliary Spring Plate No. 6	10"

PARTS LIST

No. 842

THE AUTOCAR COMPANY, ARDMORE, PENNSYLVANIA
SERVICE DEPARTMENT

BULLETIN REF.

ASSEMBLY CAB 15UBK0500 - MODEL U-2044 CHASSIS - U.S. ARMY 1940

Part No.	Req.	Description
15DL401	1	Outside Door Panel-R.H.
15DL402	1	Outside Door Panel-L.H.
15DL003	1	Door Lock Pillar-R.H.
15DL004	1	Door Lock Pillar-L.H.
15URK005	1	Door Hinge Pillar-R.H.
15URK006	1	Door Hinge Pillar-L.H.
15URK007	1	Door Top Rail-R.H.
15URK008	1	Door Top Rail-L.H.
15DL009	1	Door Bottom Rail-R.H.
15DL011	1	Door Bottom Rail-L.H.
15DL012	1	Door Belt Rail-R.H.
15DL013	1	Door Belt Rail-L.H.
15DL014	2	Door Strainer Rail
15DL218	2	Door Window Glass
15U419A	1	Door Glass Regulator-R.H.
15DL323	2	Door Regulator Channel
15DL224	3	Door Hinge-R.H.
15RL225	4	Door Dovetail-Female
15NF326	1	Door Lock-R.H.
15NF327	1	Door Lock-L.H.
15URKA328	1	Outside Door Handle-L.H.
15URK328	1	Outside Door Handle-R.H.
15NF229	2	Inside Door Handle
15DL231	2	Door Glass Felt Channel-Front
15DL032	4	Door Window Stop Block
15T233	4	Door Window Stop Block Rubber
15URK037	1	Body Lock Pillar-R.H.
15URK038	1	Body Lock Pillar-L.H.
15URK039	1	Body Hinge Pillar-R.H.
15URK041	1	Body Hinge Pillar-L.H.
15UD044A	1	Rear Window Rail-Upper
15NF248	1	Rear Window Glass
15RL251	4	Door Bumper Retainer
15DL354	1	Drip Moulding-R.H.
15UBL257	2	Roof Rear Corner Bracket
15NF264	2	Door Lock Striker Plate
15URK468	1	Rear Quarter Panel-R.H.
15URK471	1	Rear Quarter Panel-L.H.
15UBL273	1	Rear Center Panel-Top
15URK374	1	Rear Center Panel-Lower
15URK375	1	Door Sill Panel-R.H.
15URK078	1	Rear Sill
15URK580	1	Cowl Assembly
15UBL083	1	Rear Belt Rail
15URK084	1	Roof Side Rail-R.H.
15URK085	1	Roof Side Rail-L.H.

Parts List #842

Part No.	Req.	Description
15UBL086	1	Roof Rear Rail
15UBL591	1	Roof Panel
15UBK094	1	Seat Box
15UBK296	1	Sill Center Bar-R.H.
15UBK497	1	Windshield Assembly-R.H.
15URK3108	1	Tool Box Pan-L.H.
15URK2115	2	Cab Rear Panel Moulding
15SA3122A	2	Door Check Assembly
15TE2126	1	Cab Name Plate
15URK0128	1	Rear Center Pillar-R.H.
15NF2145	4	Door Bumper Rubber
15DL2146A	4	Door Dovetail-Male
15DL2147	3	Door Hinge-L.H.
15URM3152	2	Seat Back Cushion Mounting Bracket
15TF2153	2	Seat Clip-Lower
15UBM2161	1	Instrument Panel Brace
15T2171	4	Door Window Corner Reenforcement
15NF2182	2	Window Regulator Handle
15UBK4190	1	Seat Box Assembly
15NF2205	1	Rear Window Rubber Channel
15URK0218	1	Rear Cab Support Spacer Block
15URK4219	1	Inside Door Panel-R.H.
15URK4221	1	Inside Door Panel-L.H.
15UBL0224	1	Rear Window Rail-Lower
15URK2239	1	Windshield Body Bracket-Center
15URK2241	1	Windshield Body Bracket-Side
15RL2243	1	Windshield Thumb Screw
15URK2243	1	Windshield Thumb Screw
15RL2244	1	Windshield Quadrant
15URK2244	2	Windshield Quadrant Arm
15URKA2244	1	Windshield Quadrant
15URK0272	1	Rear Center Pillar-L.H.
15T2274	2	Cab Support Spring
15T2274A	2	Cab Support Spring
15UBK4284	1	Windshield Assembly-L.H.
15UBL0287	1	Ventilator Lid Side-R.H.
15UBL2292	2	Ventilator Handle-Side
15UBL2293	4	Ventilator Door Plunger Spring-Side
15UBL2294	4	Ventilator Door Plunger-Side
15UBL2295	2	Ventilator Rubber-Side
15URK2297	1	Hinge Pillar Gusset-R.H.
15UBL2302	1	Instrument Panel Bracket-R.H.
15UBL2303	1	Instrument Panel Bracket-L.H.
15NF0346	4	Door Handle & Regulator Handle Escutcheon
15DL0355	2	Door Lock Filler
15UBL0387	1	Ventilator Lid-Side-L.H.
15UBL2388A	2	Ventilator Door Plunger Bracket-Side
15URK2395	1	Roof Ventilator (Keystone)

Parts List #842

Part No.	Req.	Description
15SA2415	2	Adjusting Seat Stud
15URK2429	1	Roof Dome Light Bracket
15URK0431	1	Seat Box Post
15URK0442	1	Windshield Center Post
15SE3450	1	Rear Window Screen
15UBL0523	1	Windshield Header-R.H.
15UBL0524	1	Windshield Header-L.H.
15UBL3550	1	Ventilator Door Side Assembly-R.H.
15DL2551	1	Door Weatherstripping
15DL2612	1	Door Opening Weatherstrip
15URK0641	1	Battery Support Side-R.H.
15UBL4648	1	Front Post Upper Bracket-R.H.
15UBL4649	1	Front Post Upper Bracket-L.H.
15RL2675	1	Windshield Quadrant Screw-Female (R.H. Windshield)
15U2675	3	Windshield Regulator Bracket Screw-Female (R.H. W/S)
15RL2686	1	Windshield Quadrant Screw-Male (R.H. Windshield)
15URK2675	1	Windshield Quadrant Screw-Female
15URKA2675	3	Windshield Regulator Bracket Screw-Female
15URK2686	1	Windshield Quadrant Screw-Male
15URK3703	1	Rear Sill Corner Bracket-R.H.
15URK3705	1	Cowl Bottom Sill-R.H.
15URK3706	1	Cowl Bottom Sill-L.H.
15URK3715	1	Front Corner Post Gusset Plate-R.H.
15URK3716	1	Front Corner Post Gusset Plate-L.H.
15URK0734	1	Side Sill-R.H.
15URK0737	1	Side Sill-L.H.
15URK3756	1	Door Sill Panel-L.H.
15UD2759A	1	Rear Window Side Panel-R.H.
15UBL3764	1	Cowl Bar Gusset-R.H.
15UBL2768	1	Windshield Center Post Upper Bracket-R.H.
15URK0776	1	Seat Pan-R.H.
15URKA0776	1	Seat Pan-L.H.
15UBK0802	1	Dash Top Angle
15URK3828	1	Floorboard Side Support-R.H.
15URK0829	1	Floorboard Side Support-L.H.
15UBL0834	1	Cowl Bar-R.H.
15UBL0836	1	Cowl Bar-L.H.
15URK3837	2	Grab Handle Reenforcement Plate
15U4901A	1	Door Glass Regulator-L.H.
15UBL2908	2	Rear Cab Support Bearing Plate
15UD2914A	1	Rear Window Side Panel-L.H.
15NF2958	26	Tee Nut
15NF2959	3	Spacer
15URM0974A	1	Seat Support Angle-R.H.
15URM0975A	1	Seat Support Angle-L.H.
15URK0976	1	Seat Support Angle-Center-R.H.
15URK0977	1	Seat Support Angle-Center-L.H.

Parts List #842

Part No.	Req.	Description
15DL2985	2	Door Gusset Plate-R.H.
15DL2986	2	Door Gusset Plate-L.H.
15NF01016	2	Door Bead Filler
15UBL31019	1	Roof Rear Moulding
15UBL01028	2	Front Post Panel Filler Block
15DL21037	2	Door Window Weatherstrip
15URK01039	1	Battery Support Side-L.H.
15UBL31042	1	Cowl Bar Gusset-L.H.
15DL21045A	4	Windshield Opening Rubber
15B01024	1	Windshield Wiper Tubing-L.H. (Air)
15DLW21024	1	Windshield Wiper Tubing-R.H. (Air)
15U01024	1	Windshield Wiper Tubing-R.H. (Air)
15DLW21025	1	Windshield Wiper Tubing-L.H. (Air)
15NF21075	4	Inside Door Handle Retainer Pin
15SA01111	2	Door Check Stud
15SA01112	2	Door Check Plate
15SA01113	2	Door Check Link
15SA01114	2	Door Check Screw
15SA01115	2	Door Check Washer
15URK31270	1	Seat Support Angle Assembly-Center
15URK21374	1	Hinge Pillar Gussett-L.H.
15DL01396	1	Door Lock Regulator Plate-R.H.
15DL01397	1	Door Lock Regulator Plate-L.H.
15DL01398	2	Door Lock Regulator Plate Reenforcement
15URK21502	2	Roof Front Corner Gusset
15UBL31503	2	Roof Rear Corner Gusset
15URK01568	1	Door Hinge Pillar-R.H.
15URK01569	1	Door Hinge Pillar-L.H.
15DL01573	2	Door Glass Guide Rabbet Strip
15UBL01633	1	Fan Cover Bottom Support-L.H.
15URK01649	2	Body Lock Pillar-Lower
15URK01654	1	Roof Front Rail-R.H.
15URK01655	1	Roof Front Rail-L.H.
15UBL01658	2	Windshield Side Post Corner Filler-Upper
15UBL21661	1	Windshield Center Post Bracket-L.H.
15UBM21753	2	Instrument Panel Door Hinge-Side
15UBMA21753	1	Instrument Panel Door Hinge-Center
15UBM01754	1	Instrument Panel Door-Center
15URK01755	2	Cowl Ventilator Screen-Outside
15URKA01755	2	Cowl Ventilator Screen-Inside
15DL21777	2	Door Glass Felt Channel-Rear
15URK01784	1	Cowl Front Bottom Sill Reenforcement
15DL31804	1	Drip Moulding-L.H.
15URK31876	1	Rear Cab Sill Corner Bracket-L.H.
15UBL01903	2	Ventilator Door Hinge Side
15UBL01904	2	Ventilator Door Hinge Extension
15UBL01905	4	Ventilator Door Spring Clip
15UBL01906	2	Ventilator Door Hinge Angle-Side
15UBL31907	1	Ventilator Door Hinge Assembly-Side-R.H.
15UBL31908	1	Ventilator Door Hinge Assembly-Side-L.H.
15URK01935	1	Cowl Ventilator Screen Moulding-R.H.

Part No.	Req.	Description
15URKA01935	1	Cowl Ventilator Screen Moulding-Inside Front-R.H.
15URKB01935	1	Cowl Ventilator Screen Moulding-Inside Rear-R.H.
15URK01936	1	Cowl Ventilator Screen Moulding-L.H.
15URKA01936	1	Cowl Ventilator Screen Moulding-Inside Front-L.H.
15URKB01936	1	Cowl Ventilator Screen Moulding-Inside Rear-L.H.
15UBK01983	1	Sill Center Bar-L.H.
15U22013A	1	Windshield Regulator Bracket
15UBL32040	1	Ventilator Door Side Assembly-L.H.
15U22056	1	Windshield Regulator Bracket Pin
15URM02103A	1	Seat Box Reenforcement
15URK02108	1	Tool Box Pan Insert
15URK02123	2	Body Lock Pillar-Upper
15DL22131	308"	Door Weatherstrip Rubber
15URK22133	2	Seat Box Corner Gusset
15UBL22134	1	Windshield Body Bracket Clip
15UBM22135	1	Windshield Body Bracket Hinge
15DL22139	4	Door Window Glass Roller
15U22156	2	Instrument Panel Door Knob
15UBM02157	2	Instrument Panel Door-Side
15UBL02301	1	Radiator Filler Cap Door
15UBL22302	1	Radiator Filler Cap Door Hinge
15UBL02303	1	Radiator Filler Cap Door Spring Clip
15UBL32310	1	Radiator Filler Cap Door Assembly
15URM22317	4	Seat Angle Retainer
15UBL02325	1	Radiator Filler Cap Door Hinge Extension
15UBL32326	1	Radiator Filler Cap Door Hinge Assembly
15UBL02327	1	Radiator Filler Cap Door Hinge Angle
15UBL22328	1	Radiator Filler Cap Door Plunger
15UBL22329	1	Radiator Filler Cap Door Plunger Spring
15UBL22344	1	Radiator Filler Cap Door Rubber
15UBL32347	1	Radiator Filler Shield Assembly
15UBK42348	1	Radiator Top Shield Assembly
15UBK32351	1	L.H. Floorboard Support Front Angle
15UBK32352	1	L.H. Floorboard Support Angle Assembly-Center
15UBK02355	1	L.H. Floorboard Support Angle-Center
15UBL02356	1	Cowl Bar Filler End
15UBK42366	1	Radiator Top Shield
15UBL22367	2	Radiator Top Shield Felt Retainer
15UBL22368	2	Radiator Filler Shield Felt Retainer
15UBL02373	2	Windshield Header Filler Strip
15UBK02375	1	L.H. Floorboard Support Angle Brace
15UBKA22374	1	Fan Cover Side Felt Retainer-Upper
15UBM02381	1	Glove Compartment Body
15UBM02382	1	Glove Compartment End-R.H.
15UBM02383	1	Glove Compartment End-L.H.
15UBL02437	4	Front Corner Post Bracket Hole Plug
15UBL22479	1	Windshield Center Post Shim
15UBM22623	2	Instrument Panel Compression Door Spring
15URK22632	1	Seat Support Angle Bracket-Center
15URK22633	1	Seat Support Angle Bracket-Side R.H.

Parts List #842

Part No.	Req.	Description
15URK22638	1	Seat Support Angle Bracket-Side-L.H.
15UBL02953	1	Windshield Defroster Air Duct Body-R.H.
15UBLA02953	1	Windshield Defroster Air Duct Body-L.H.
15UBL02954	1	Windshield Defroster Air Duct Bracket-Side
15UBL02955	1	Windshield Defroster Air Duct Bracket-End
15URK22976	2	Door Check Bushing
15URK22986	1	Roof Ventilator Gasket
15URK22987	1	Cowl Ventilator Screen Assembly-Outside-R.H.
15URKA22987	1	Cowl Ventilator Screen Assembly-Inside-R.H.
15URK22988	1	Cowl Ventilator Screen Assembly-Outside-L.H.
15URKA22988	1	Cowl Ventilator Screen Assembly-Inside-L.H.
15URK02989	2	Cowl Ventilator Screen Moulding Clip
15UBK32991	1	Fan Cover Side Extension Assembly-R.H.
15UBK02992	1	Fan Cover Side Extension-Front
15UBK02993	1	Fan Cover Side Extension Body-R.H.
16URK3165	2	Grab Handle
16Y0171	25-1/2"	Anti Rattle Tape
16URK2181	1	Ventilator Door Wind Split
16U2158	1	Windshield Wiper "T" Fitting (Air)
16UBL4249	1	Headlamp Bracket-R.H.
16UBL4251	1	Headlamp Bracket-L.H.
16URK3341	1	Seat Spring-R.H.
16URM3342	1	Seat Back Spring
16UBK0421	1	Fan Cover-Center
16URK0441	1	Cowl Panel-Bare
16UBL2443A	1	Starting Switch Bracket Assembly
16U2565	1	Oil Gauge Flexible Tubing
16URK2646	1	Fire Extinguisher
16URK3713	1	Battery Shield Side
16SA2724	4	Rubber Grommet
16URK2724	2	Headlamp Cable Grommet
16UBK3768	2	Rear View Mirror Assembly
16URK2723	1	Rear Cab Support Top Bearing Plate
16URK4823	2	Windshield Wiper (Air)
16UBK4828	1	Fan Cover Assembly
16URK3990	1	Seat Back Cushion Assembly
16URK31061	1	Seat Spring-R.H.
16UBL21096	43"	Floorboard Rubber Mat
16URK31110	1	Battery Support Assembly
16ZURM01154	1	Seat Back Frame Rail-Lower
16ZURM01161	2	Seat Back Frame Rail-Side
16ZURMA01161	2	Seat Back Frame Rail-Top Side
16DFL21081	2	Instrument Panel Spacer (Air)
16UBK41188	1	Gear Shift Lever Rubber Boot
16UBL41188	1	Gear Shift Lever Rubber Boot
16UBL21244	1	Foot Dimmer Switch Bracket
16ZUD21341	2	Battery Cable Bracket
16URKA01396	2	Battery Hold Down Frame Side
16D21404	1	Steering Column Finish Rubber

Parts List #842

Part No.	Req.	Description
16URK21444	1	Horn Relay Bracket
16URK21447A	1	Fuse Block Bracket
16UBL01449C	1	Starting Switch Bracket
16ZTEA01397A	2	Battery Hold Down Frame End
16UBK31252	1	Fan Cover Side Filler-R.H.
16URK31460	1	Battery Hold Down Frame Assembly
16RLA21515	1	Crank Hole Opening Reenforcement
16RL21758	1	Headlight Switch Knob
16UBK21833	1	Cable Housing-Upper
16UBKA21833	1	Cable Housing-Lower
16UG21896	1	Shift Plate (Main Trans.)
16UL21896A	1	Shift Plate (Transfer Case)
16UBK31899	1	Fan Cover Center Open Lid
16UBK41911	1	Fan Cover Assembly-Side-R.H.
16UBL31912	1	Fan Cover Assembly-Side-L.H.
16UBK01913	1	Fan Cover Side-R.H.
16UBL01914	1	Fan Cover Side-L.H.
16UBL22145	1	Choke Control Bracket
16UBL22154	1	Starting Motor Cable Boot
16UBL22155	1	Starting Motor Cable Boot Clamp
16UBK02181	1	Fan Cover Center Filler
16URK32220	1	Seat Cushion Assembly-R.H.
16URK32230	1	Seat Cushion Assembly-L.H.
16UK22285	1	Caution Plate
16UKA22285	1	Name Plate
16URK22330	1	Windshield Wiper Tubing Assembly (Air)
16ZURM02453	2	Seat Back Frame Rail-Center Top
16URK22573	1	Toggle Switch Mounting Bracket
16UG22618	1	Rear Cab Support Filler Block Assembly
16UEK02776	1	Chain Box Body
16UBK02777	1	Chain Box Side-R.H.
16UBK02778	1	Chain Box Side-L.H.
16UBK02779	1	Chain Box Door
16UBK02781	1	Chain Box Hinge
16ZRLA02782	1	Chain Box Door Lock Block
16HF22786	3	Tool Bag Draw Rope
16HF02793	1	Tool Bag Assembly
16HF22794	1	Key Container
16URK22587	1	Fire Extinguisher Mounting Bracket

PARTS LIST

No. 849

THE AUTOCAR COMPANY, ARDMORE, PENNSYLVANIA
SERVICE DEPARTMENT

BULLETIN REF.

ASSEMBLY   FENDERS AND SHEET METAL - MODEL U-2044 - U.S.A. 1940

Part No.	Req.	Description
16UBK5950	1	Engine Hood Assembly
16UBL22106	1	Engine Hood Insulation
16UBL22107	1	Engine Hood Insulation Cement
16UBK41911	1	Fan Cover Assembly - R.H.
16UBL31912	1	Fan Cover Assembly - L.H.
14URK052A	1	Fender Assembly - R.H.
14URK053A	1	Fender Assembly - L.H.
14URK4101	1	Fender Skirt Assembly - R.H.
14URK4102	1	Fender Skirt Assembly - L.H.
14TE0125	1	Fender Skirt Binder (91 inches)
14URK321	1	Splash Guard - R.H.
14URK322	1	Splash Guard - L.H.
14URK343	1	Running Board - R.H.
14URK344	1	Running Board - L.H.
14URK361A	2	Fender Step - Top
14URKA361A	2	Fender Step Assembly - Center
14URK478	1	Running Board Brace - R.H.
14URKA478	1	Running Board Brace - L.H.

SERVICE BULLETIN No. 693

THE AUTOCAR COMPANY, ARDMORE, PENNSYLVANIA
SERVICE DEPARTMENT

SUBJECT - ELECTRICAL SYSTEM WITH STEP-VOLTAGE CONTROL

The complete electrical system including generator, starter, battery and lights is illustrated diagrammatically on the inside spread of this leaflet. Note that a single wire system with grounded return is used.

GENERATOR (Step-Voltage Type)

With this system the charging rate is automatically controlled so that a high charging rate is obtained when the load is heavy or when the battery is in a discharged condition and a gradually decreasing rate as the load is reduced or as the battery becomes fully charged. This system operates automatically to prevent overcharging or overheating of the battery and insures most efficient operation.

Lubrication -- The voltage regulated generator is fitted with ball bearings at both ends. Place 3-4 drops of light engine oil in the oiler on the commutator end every 1000 miles. The drive end bearing is lubricated automatically from the timing gear case.

Adjustments -- Do not undertake any adjustments on generator or regulator unless proper testing equipment is available. Changes in third brush setting or in the regulator adjustments may increase the charging rate beyond the capacity of these units and cause them to burn out. Refer to Autocar Service Stations for inspection and adjustment of these units.

With the control unit and generator properly set it will not be necessary to check or readjust except at intervals of 10,000 miles.

General Care -- The commutator and brushes should be kept clean as good brush contact is essential to good generator performance. Clean and smooth the commutator with a small piece of fine sandpaper if discolored. DO NOT USE EMERY CLOTH.

The brushes must move freely in their holders. Remove brush dust from brushes and holders to prevent sticking of brushes.

Excessive arcing at the brushes may result from improper brush spring pressure, sticking brushes, extremely short brushes, rough or eccentric commutator or overloading of the generator. Any or all such conditions should be corrected to avoid damage to the brush holders or bracket and melting of solder on the commutator which may cause an open circuit in the armature windings.

STEP-VOLTAGE CONTROL UNIT

The Step-Voltage Control, composed of a voltage control unit and cut-out relay, is mounted on the generator. This unit operates to decrease the maximum generator output when the battery reaches a fully charged condition.

If trouble develops do not disturb the control unit until other possible causes are checked, particularly with respect to mechanical condition of generator. Adjustment or repairs to the control unit should be attempted only where complete facilities are available for handling and testing such work.

BATTERY

The battery and its compartment should be kept clean and dry. The battery terminals should also be kept clean, tight and coated with grease to prevent corrosion.

State of Charge -- The state of charge can be determined by use of battery hydrometer. Hydrometer readings of 1.275-1.285 (Specific Gravity) indicate full

⟵ See Wiring Diagram in Fold   Text continued on page 2 ⟶

BATTERY - Continued

charge. In service the battery should not be allowed to drop under 1,200 at which point the battery should be charged from an outside source.

Freezing Point -- In cold weather the battery should not be permitted to reach a discharged state as it is subject to freezing when in this condition with the electrolyte under 1180 specific gravity.

Adding Water -- Distilled or other approved water must be regularly added to keep the level of the electrolyte above the top of the separators. Under average conditions the level of the liquid should be inspected every two weeks. The level should be maintained at a height 3/8" above the separators.

STARTING MOTOR

The starting motor operates through a reduction gear built into the motor housing and engages and disengages the flywheel gear through operation of a Bendix drive.

Lubrication -- The starting motor armature shaft is mounted on grey iron bearings which are lubricated by means of oil cups at each end of the unit. Place 3-4 drops of light engine oil in these cups every 1000 miles or every week. The pinion housing is fitted with an oil-less bearing. Remove grease plug in reduction gear housing and repack gears with graphite grease every six months.

Bendix Drive -- The Bendix Drive assembly is entirely automatic and requires no oiling or attention. Failure of the starter mechanism to engage the flywheel gear may be caused by grease or other gummy deposits on the Bendix mechanism. If trouble of this kind is encountered, take out the starter and wash off the driving mechanism.

General Care -- It is a very good policy to remove the cover band on the starting motor and inspect the brushes and commutator at least once a year. The brushes should be replaced if badly worn and commutator cleaned if dirty or rough. Use a fine sandpaper.

Failure of the starter to turn over the engine or sluggish action of the starter may be the result of either a loose or corroded battery terminal, a discharged battery, a defective starting switch, short or open circuit between the battery, starting switch or starting motor.

LIGHTS

Headlights -- The headlamps embody the depressible beam feature which involves the use of a double filament bulb with one filament located on the exact center of the reflector and the other slightly above the center.

The operation of the depressible beam is effected by means of a foot switch located on the toe-board to the left of the clutch pedal.

The lamps have fixed focus but can be adjusted directionally by loosening the supporting bolt under the lamp and shifting the position of the lamp on the supporting bracket.

Tail and Stop Light -- A combination tail and stop lamp is used. The stop light is controlled by means of a switch connected to the brake pedal.

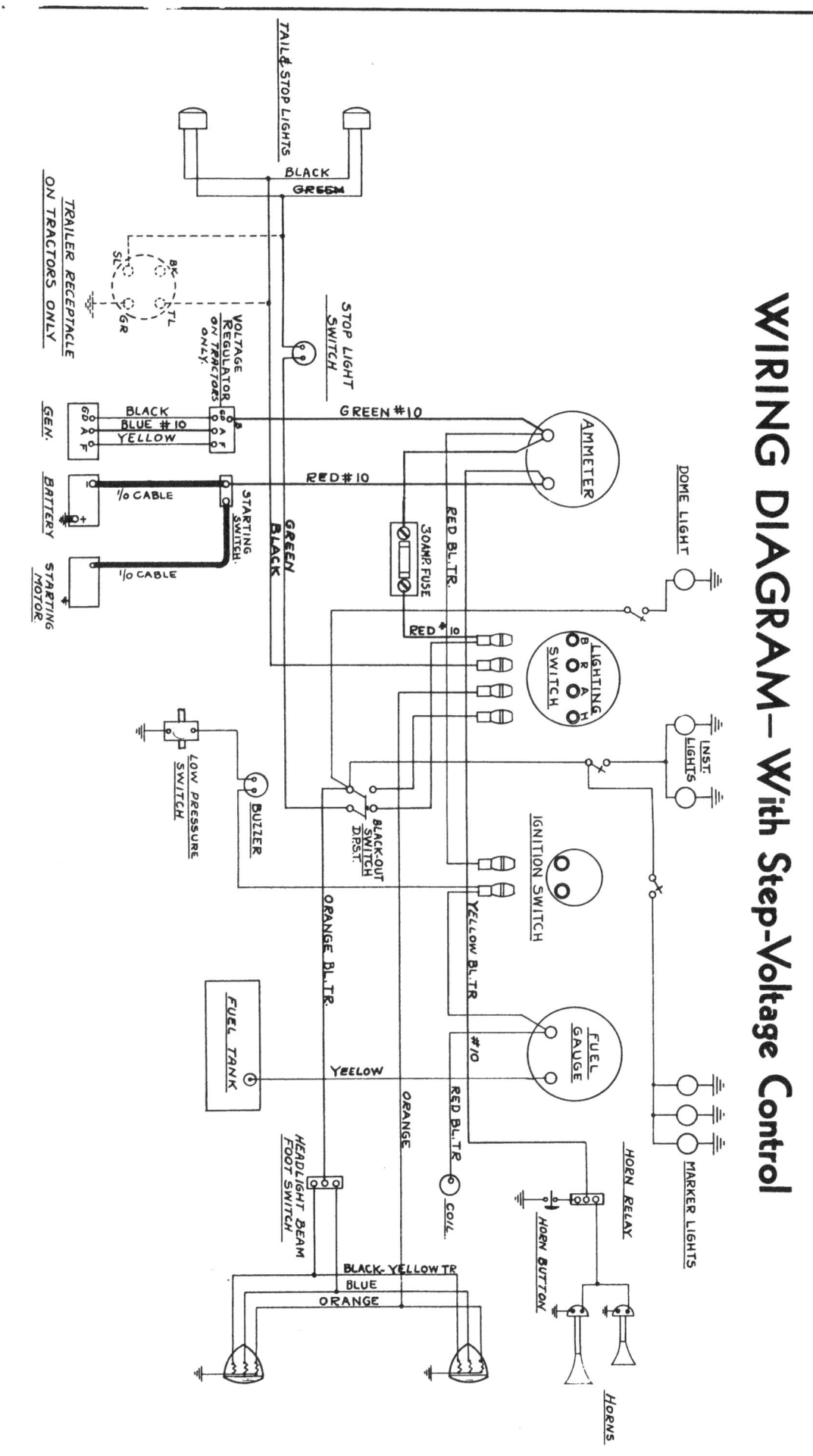

PARTS LIST

No. 843

THE AUTOCAR COMPANY, ARDMORE, PENNSYLVANIA
SERVICE DEPARTMENT

BULLETIN REF.

ASSEMBLY ELECTRICAL UNITS AND WIRING - MODEL U-2044 CHASSIS-U.S. ARMY 1940

Part No.	Req.	Description	Reference Number
		MAIN UNITS	
16BLF0650	1	Generator & Gear Assembly	
16BLF0410	1	Generator Assembly (Delco Remy)	DR-1102404
2BL0495	1	Generator Gear (Hercules)	40266-B
16BLG0620	1	Voltage Regulator (Delco Remy)	DR-5540
16BLA0560	1	Starting Motor (Delco Remy)	DR-720T
16SCM0880	1	Bendix Drive Assembly (Eclipse Machine Co.)	R11X-13
16SCM02004	1	Bendix Sleeve & Shaft Assembly	R11X-13SA
16A0595	1	Bendix Drive Head	SR10-105
16T0596	1	Bendix Spring Support Clip	R11-21
16A0539	2	Bendix Spring Screw Lock Washer	R12-8
16A0593	1	Bendix Head Spring Screw	R11-109X
16A0592	1	Bendix Shaft Spring Screw	R11-7X
16A0594	1	Bendix Service Sleeve	R11-112R
16T0538	1	Bendix Drive Spring	R11-6X
16ZRM0440	1	Battery 6-V (Exide)	#3LXRH17-1R
		SWITCH GROUP	
16BL2680B	1	Starting Switch (Auto-Lite)	SW-4002
16RL2420B	1	Ignition Switch (H.A. Douglass)	#2980
16RL2699A	1	Foot Dimmer Switch (H.A. Douglass)	#5530
16NF01241	1	Toggle Switch-Single Pole (Arrow, Hart & Hagemen)	#8963
16URK21241	1	Blackout Switch (Cutler-Hammer)	#8620-K3
16RL21755	1	3-Position Light Switch (H.A. Douglass)	#5399
25NT2130	1	Stop Light Switch (Westinghouse)	215537
		INSTRUMENTS AND GAUGE GROUP (STEWART WARNER)	
16RL32660B	1	Instrument Cluster Assembly	#93499
16RL02364	1	Instrument Cluster Case, Bezel & Glass Assem.	93564
16RL02362	1	Instrument Cluster Case & Window Assem.	74096
16RL02363	2	Instrument Cluster Mounting Bracket	74074
16RL0630	1	Ammeter	90707
16RL0775	1	Gas Gauge	93489
16RL02040	1	Heat Indicator	93619
16RL0564	1	Oil Gauge	96808
16UB3760	1	Speedometer Assembly (Stewart Warner)	#584AM
16K9774	1	Fuse (30 Amp. - 4 A.G.)	
16SA2773	1	Fuse Block Assem.	
16DFLB3786-15"	1	Gas Tank Gauge Unit (In tank)	
16URK32740	1	Electric Horn Assem. (Dual.) (E.A. Lab.)	E.A. 44
	1	Electric Horn Relay (E.A. Lab.)	E.A. 93

-2-   Parts List #843

Part No.	Req.	Description	Reference Number
		HEADLAMP GROUP (Guide Lamp Company)	
16UBK4460	2	Headlamp Assembly	675Z
16UBK0988	2	Headlamp Body Assembly	5931708
16UBK0365	2	Headlamp Moulding Assembly	5931167
16HF0366	2	Headlamp Moulding Screw	922690
16URK0992	2	Headlamp Wire Assembly	5931709
16URK01079	8	Headlamp Rubber Sleeve	804463
16URK0265	8	Headlamp Eyelet Terminal	1867662
16URK0915	2	Headlamp Douglass Parking Wire Terminal	1878192
16URK0261	2	Headlamp Female Sleeve	502793
16URK0579	2	Headlamp Wire Cover Loom	5931595
16URK0274	2	Headlamp Loom Clip	5930502
16UBK0286	2	Headlamp Retaining Ring-Prime	5931243
16URK0759	6	Headlamp Retaining Ring Screw	924552
16URK0336	2	Headlamp Gasket	5930408
16URK0367	2	Headlamp Lens	5930411
16URK0724	2	Headlamp Grommet	5930487
16URK0553	2	Headlamp Reflector	5930400
16URK0274A	6	Headlamp Reflector Clip	5930413
		PARKING LAMP (Part of Headlamp Assembly)	
16UBK02627	2	Parking Lamp Assembly	88L
16UBK02628	2	Parking Lamp Body Assembly-Prime	5930477
16URK0915	2	Parking Lamp Douglass Terminal	1878192
16UBK02629	2	Parking Lamp Door Assembly (Less Lens) Prime	5931176
16URK02631	2	Parking Lamp Lens-Frosted	923277
16URK02632	2	Parking Lamp Gasket	922758
16URK02633	2	Parking Lamp Door Screw	922446
16URK02634	2	Parking Lamp Mounting Gasket	5930478
16URK02635	2	Parking Lamp Mounting Lock Washer	138485
16URK02636	2	Parking Lamp Mounting Bolt Nut	121917
		DOME LIGHT GROUP	
16URK21054	1	Roof Dome Light	
		BULB GROUP	
16SA0151	2	Dome Light Bulb 6-8V - 15CP	
16CK2472	1	Dash Light Bulb 6-8V - 3CP	
16A0472	1	Dash Light Bulb 6-8V (Air Gauge)	
16J0729	2	Headlamp Bulb 6-8V - 32/32CP	
16A0472	2	Parking Lamp Bulb 6-8V	
		CABLE ASSEMBLIES	
16AK950	1	Battery Ground Cable Assembly (Positive)	
16URK2480	1	Headlamp Cable Assembly-R.H.	
16URK2490	1	Headlamp Cable Assembly-L.H.	

-3-   Parts List #843

Part No.	Req.	Description	Reference Number
		CABLE ASSEMBLIES (Continued)	
16AG3580	1	Starting Switch to Motor Cable Assembly	
16AA9590	1	Battery to Starting Switch Cable Assembly	
16S9623	1	Jumper Wire Assembly	
16BG9623	1	Jumper Wire Assembly	
16BF9623	1	Jumper Wire Assembly	
16BD9623	1	Jumper Wire Assembly	
16BE9623	1	Jumper Wire Assembly	
16D9623	1	Jumper Wire Assembly	
16UBK3510	1	Tail Lamp Cable Assembly	
16UBK0802	1	Wiring Unit Assembly - Complete	
16UBK02350	1	Generator Cable Assembly	
16UBM02470B	1	Dimmer Switch Cable Assembly	
		WIRES & MISCELLANEOUS	
16CE0109A	110"	Double Conductor Wire	
16A0268	87"	Battery Cable	
16A0273	11"	Loom	
16SA0273	352"	Loom - 1/2"	
16TE0273	73"	Loom - 5/8"	
16Y0273	304"	Loom - 3/8"	
16A0281	446"	Black Wire #14 Gauge	
16D0281	197"	Black Tracer Wire #14 Gauge	
16D0282	141"	Red Tracer Wire #14 Gauge	
16TE0282	155"	Red Wire #10 Gauge	
16A0282	430"	Red Wire #14 Gauge	
16D0873	65"	Orange Tracer Wire #14 Gauge	
16C0873	162"	Orange Wire #14 Gauge	
16SA0873	393"	Green Wire #14 Gauge	
16D0874	115"	Yellow Tracer Wire #14 Gauge	
16SA0874	193"	Yellow Tracer Wire #14 Gauge	
16SA0875	199"	Blue Wire #14 Gauge	
16TE0875	80"	Wire	
16E0874	130"	Yellow Wire (Black Tracer) #10 Gauge	
16UBL2252	1	Battery Terminal (Positive)	
16SA2253	1	Battery Terminal (Negative)	
16A0277	46	Rubber Boot	
16RL2265	8	Terminal	
16A2487	8	Cable Clamp	
16C2499	1	Terminal - 3/8"	
16SA2499	3	Terminal	
16URK2724	2	Headlamp Cable Grommet (Western Rubber Company)	
16URK2911	2	Battery Clamp Screw	
16UBM2933	1	Cable Clamp	
16RL0951	1	Dash Light Socket Assem.	
16ZDFL2951	1	Dash Light Socket Assem. (Air Gauge)	
16URK01416	4	Terminal	
16N21416	1	Terminal	
16RL21755	8	Wire Connection (Douglass)	
16SA2724	4	Rubber Grommet	

PARTS LIST

No. 843 Suppleme

THE AUTOCAR COMPANY, ARDMORE, PENNSYLVANIA
SERVICE DEPARTMENT

BULLETIN REF.

ASSEMBLY   16BLF0410-GENERATOR ASSEMBLY - DELCO-REMY MODEL 1102404

Description	Delco-Remy Number
Generator Armature	1856861
Generator Field Coil-R.H.	1838577
Generator Field Coil-L.H.	1845109
Generator Pole Shoe	816323
Generator Commutator End Frame	1838158
Generator Drive End Frame	822643
Generator Ball Bearing - C.E.	903203
Generator Ball Bearing - D.E.	903204
Generator Ball Bearing Retainer Plate-D.E.	820707
Generator Ball Bearing Retainer Plate-D.E.	820709
Generator Fan	1864264
Generator Main Brush Plate	1853020
Generator Third Brush Plate	1850763
Generator Main Brush	1860344
Generator Third Brush	1850768
Generator Main Brush Spring	1850760
Generator Third Brush Spring	1850767
Generator Brush Holder	1850759
Generator Third Brush Plate Clamp	1857187
Generator Third Brush Plate Clamp Spring	1850771
Generator Brush to Relay Lead	821449
Generator Thru Bolt	812291
Generator Commutator Cover Band	1861502
Generator Bearing Retainer Screw Lock Washer	106497
Generator Brush Lead Screw Lock Washer	802730
Generator Control Unit Screw Lock Washer	106496
Generator End Cover Screw Lock Washer-C.E.	138479
Generator Shaft Nut Lock Washer	804000
Generator Third Brush Clamp Screw Lock Washer	106496
Generator Thru Bolt Lock Washer	108579
Generator Shaft Nut	806915
Generator Bearing Retainer Screw	1866970
Generator Brush Lead Screw	141542
Generator Main Brush Plate Screw	1868330
Generator Control Unit Screw	132900
Generator Pole Shoe Screw	828675
Generator Upper Third Brush Clamp Screw	115434
Generator Brush Arm Spacer Washer	1857412
Generator Felt Washer - D.E.	820706
Generator Shaft Nut Washer	30205
Generator Inside Spacer Washer - D.E.	822644
Generator Relay Lead Bushing	812891
Generator Inside Collar - C.E.	811681
Generator Inside Collar - D.E.	822644
Generator Outside Collar - D.E.	1858603
Generator Outside Collar - D.E.	1836194

Page 114

Parts List No. 843
Supplement

Description	Delco-Remy Number
Generator Dowel Pin - C.E.	809062
Generator Dowel Pin - D.E.	809593
Generator Bearing Retainer Gasket - D.E.	820708
Generator Oiler - D.E.	125609
Generator Oiler - D.E.	113702
Generator Control Unit Terminal Clip	811884
Generator Brush Terminal Clip	1857107
Generator Gear Key	124548

NOTE: C.E. denotes Commutator End
D.E. denotes Drive End

PARTS LIST

No. 843 Supplement

THE AUTOCAR COMPANY, ARDMORE, PENNSYLVANIA
SERVICE DEPARTMENT

BULLETIN REF.

ASSEMBLY 16BLA0560 - STARTING MOTOR ASSEMBLY - DELCO-REMY MODEL 720-T

Description	Delco-Remy Number
Starting Motor Frame & Field Assembly	817790
Starting Motor Field Coil - R.H.	810627
Starting Motor Field Coil - L.H.	810626
Starting Motor Commutator End Frame	815839
Starting Motor Bendix Housing	811299
Starting Motor Bendix Housing Bushing	811230
Starting Motor Brush & Field Connector Lead	810586
Starting Motor Armature	818002
Starting Motor Brushes	811553
Starting Motor Ground Brush Lead	813554
Starting Motor Field Brush Lead	811450
Starting Motor Brush Spring	813521
Starting Motor Brush Holder	810226
Starting Motor Thru Bolt	809053
Starting Motor Commutator Cover Band	817114
Starting Motor Brush to Field Screw Lock Washer	141552
Starting Motor Brush Ground Lead Screw Lock Washer	141551
Starting Motor Field Terminal Lock Washer	142248
Starting Motor Thru Bolt Lock Washer	141553
Starting Motor Field Terminal Nut	805258
Starting Motor Field Terminal Nut	114503
Starting Motor Oil Wick - C.E.	802691
Starting Motor Oil Wick & Spring - C.E.	809591
Starting Motor Brush Holder Pin	817313
Starting Motor Brush Holder Stop Pin	817314
Starting Motor Brush Holder Pin & Insulation	812016
Starting Motor Brush Holder Pin & Insulation	812015
Starting Motor Dowel Pin - D.E.	809593
Starting Motor Dowel Pin - 1/8 - D.E. & C.E.	809062
Starting Motor Brush to Holder Screw	115903
Starting Motor Brush Lead to Field Screw	135616
Starting Motor Brush Ground Screw	114935
Starting Motor Pole Piece Screw - 5/16-18	802790
Starting Motor Pole Piece Screw - 3/8-24	828675
Starting Motor Field Terminal Stud	826938
Starting Motor Field Terminal Stud Washer	831688
Starting Motor Field Terminal Stud Washer	809051
Starting Motor Field Terminal Stud Washer	817077
Starting Motor Armature Shaft Spacer Washer	811388
Starting Motor Armature Shaft Spacer Washer	833602
Starting Motor Bendix Shaft Spacer Washer	812496
Starting Motor Bendix Shaft Spacer Washer	812664
Starting Motor Brush Lead Terminal Clip	811451
Starting Motor Brush Lead Terminal Clip	816453
Starting Motor Brush Lead Terminal Clip	808933
Starting Motor Dust Cap - C.E.	810819
Starting Motor Field Terminal Wedge	810824

Parts List No. 843
Supplement

Description	Delco-Remy Number
Starting Motor Oiler	809595
Starting Motor Oil Well Plug - C.E.	103884
Starting Motor Pole Piece	810601
Starting Motor Woodruff Key No. 6	106750
Starting Motor Brush Connector Lead	819362

NOTE: C.E. Denotes Commutator End
D.E. Denotes Drive End

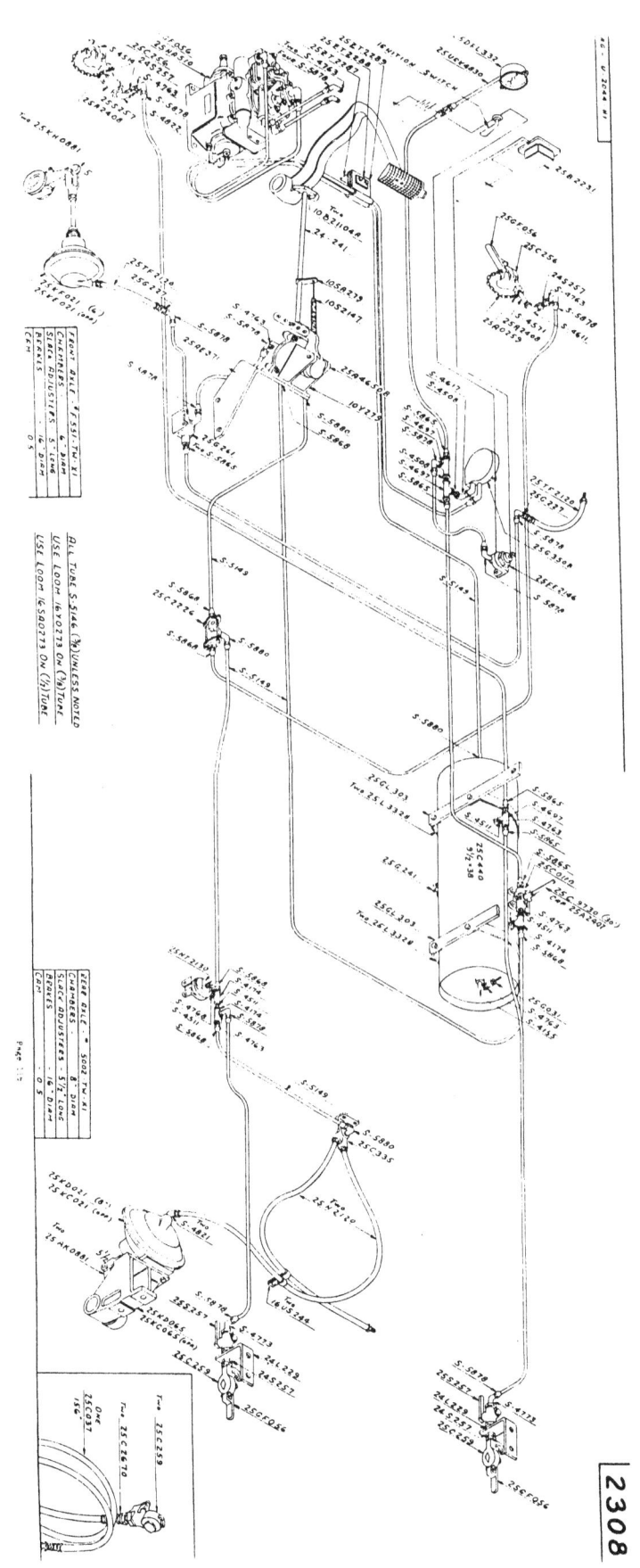

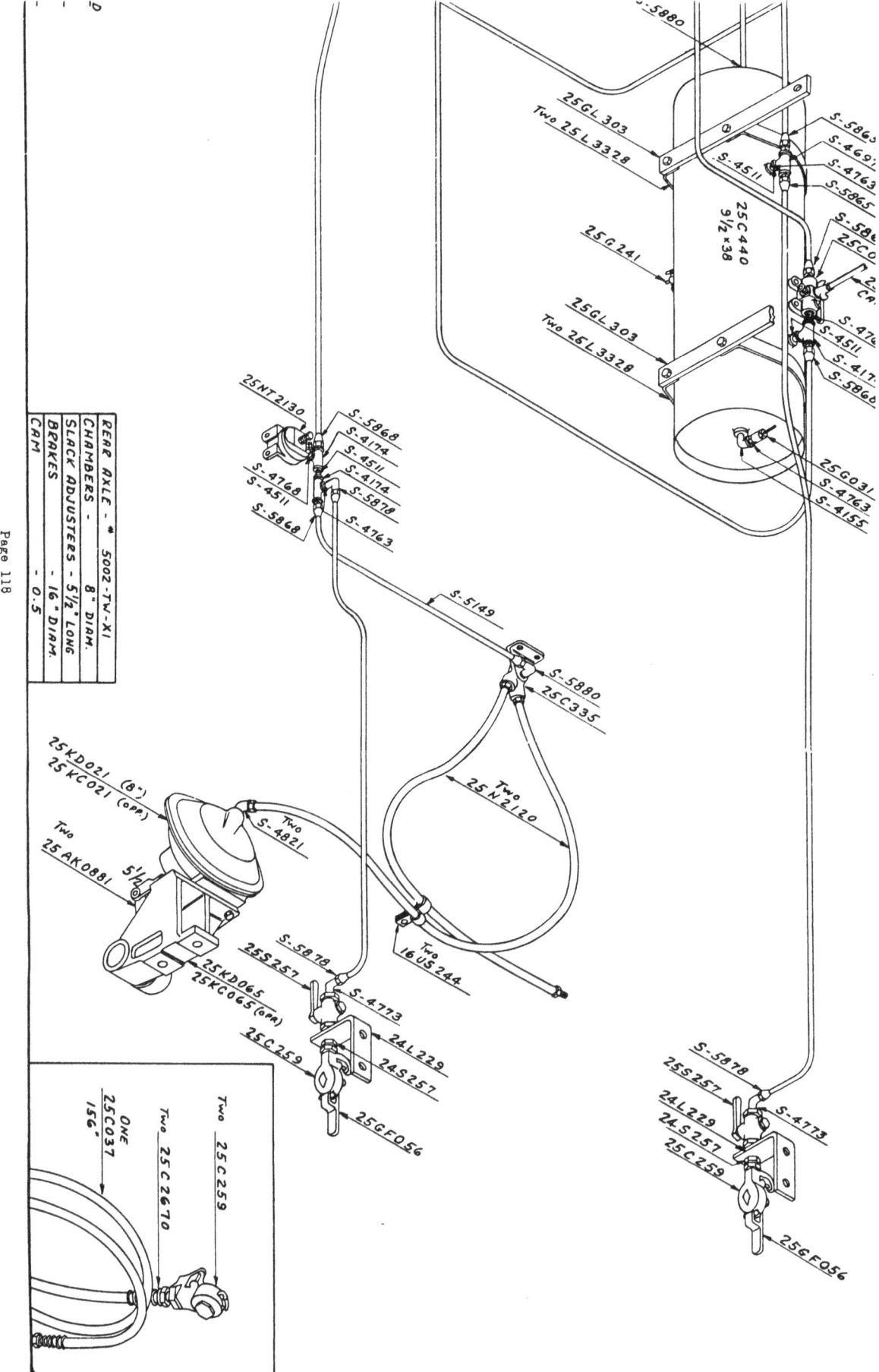

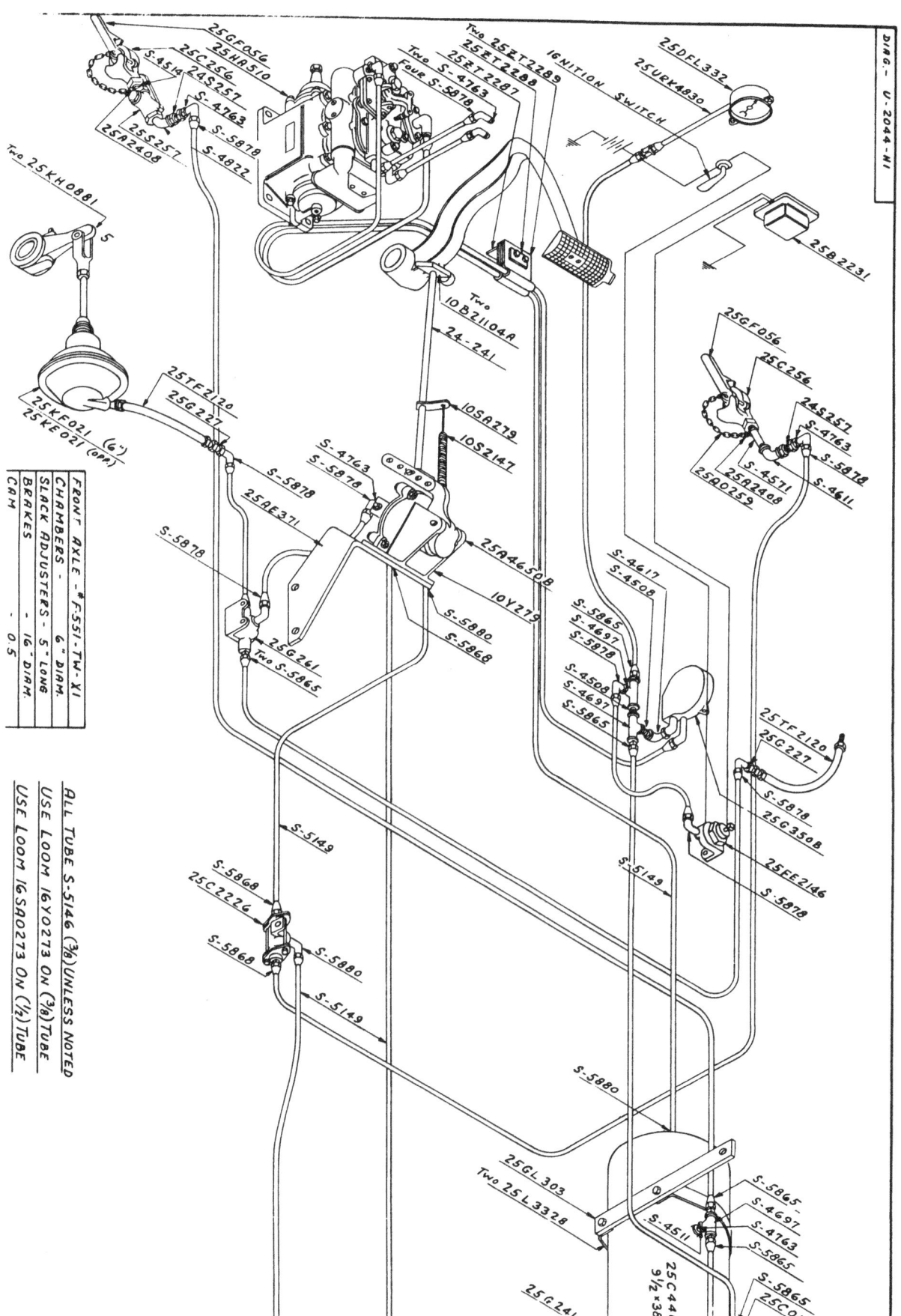

PARTS LIST

No. 858

THE AUTOCAR COMPANY, ARDMORE, PENNSYLVANIA
SERVICE DEPARTMENT

BULLETIN REF.

ASSEMBLY AIR BRAKE EQUIPMENT - MODEL U-2044 CHASSIS - U.S. ARMY 1940

Reference Brake Sketch #2308

Part No.	Req.	Description	Reference Number
10SA279	1	Accelerator Release Spring Link	
10Y279	1	Brake Release Spring Link	
10S2147	1	Retracting Spring	
10B21104A	2	Offset Clevis Pin	
16US244	2	Clip	
24L229	2	Pipe Bracket	
24S257	4	Frame Nipple	
24-241-15"	1	Brake Rod	
25GL303	2	Air Tank Bracket	
25HA510	1	Air Compressor (Bendix-Westinghouse)	220148
25KC021	1	R.H. Rear Brake Chamber (8") (Bendix-West.)	216475
25KD021	1	L.H. Rear Brake Chamber (8") (Bendix-West.)	216476
25KE021	1	R.H. Front Brake Chamber (6") (Bendix-Westinghouse)	220127
25KF021	1	L.H. Front Brake Chamber (6")(Bendix-West.)	220128
25G227	2	Frame Tube	
25G031	1	Safety Valve (Bendix-Westinghouse)	205105
25DFL332	1	Air Pressure Gauge Assem. (Stewart-Warner)	100248
25C335	1	Air Hose Tee	
25C037	1	Hose Assembly (156") (Bendix-Westinghouse)	101-M
25C440	1	Air Tank (9-1/2" x 38")	
25G241	1	1/4" Reservoir Drain Cock (Bendix-West.)	215310
25G350B	1	Governor (Bendix-Westinghouse)	215995
25C256	2	Dummy Hose Coupling (Bendix-Westinghouse)	212227
25GF056	4	Dummy Hose Coupling (Bendix-Westinghouse)	216581
25S257	3	Shut Off Cock (1/2" x 1/2") (Imperial Brass Mfg. Co.)	#210
25C259	4	Hose Coupline (1/2") (Bendix-Westinghouse)	217766
25G261	1	Quick Release Valve (Bendix-Westinghouse)	205000
25KC065	1	R.H. Brake Chamber Bracket	
25KD065	1	L.H. Brake Chamber Bracket	
25AE371	1	Application Valve Bracket	
25C0110	1	Tire Inflating Valve Assem. (Bendix-West.)	205852
25N2120	2	Air Brake Hose Assem. (34-1/2") (Bendix-Westinghouse)	215024
25TF2120	2	Air Brake Hose Assem. (19-1/2") (Bendix-Westinghouse)	205890
25NT2130	1	Stop Light Switch (Bendix-Westinghouse)	215537
25FE2146	1	Low Pressure Indicator (Bendix-Westinghouse)	215186
25C2226	1	Air Double Check Valve Assem. (Bendix-Westinghouse)	217698
25B2231	1	6-V Buzzer (Electric Service & Supply Co.)	11120
25A0259	1	Chain (12" - #3 Knotted Galvanized)	
25ZT2287	1	Air Line Rubber Mounting Bracket	

Reference Brake Sketch #2308   -2-   Parts List #858

Part No.	Req.	Description	Reference Number
25ZT2288	1	Air Line Rubber Mounting Bracket Plate	
25ZT2289	1	Air Line Mounting Rubber	
25L3328	4	Air Tank Support Strap	
25A2408	2	Link	
25A4650B	1	Air Application Valve & Stop Assembly	
25C2670	2	Flexible Hose Connector (Bendix-Westinghouse)	215535
25C9730	1	Tire Inflation Hose Assembly (30")	
25URK4830	1	Air Windshield Wiper Tubing Assembly	
25AK0881	2	Rear Brake Slack Adjuster (Bendix-West.)	215530
25KH0881	2	Front Brake Slack Adjuster (Bendix-West.)	215091
S-4155	1	Ell (3/8 x 90° Brass)	
S-4174	3	Tee (3/8" Brass)	
S-4508	2	Nipple (1/4" Brass-Close)	
S-4511	3	Nipple (3/8" Brass-Close)	
S-4514	1	Nipple (1/2" Brass-Close)	
S-4571	1	Nipple (1/2" x 3" Brass)	
S-4611	1	Ell (1/2" x 45° Brass)	
S-4617	1	Ell (1/4" x 90° Brass)	
S-4697	3	Tee (1/4" Brass)	
25A2407	1	Cap	
S-4763	9	Bushing (3/8" x 1/4" Brass)	
S-4768	1	Bushing (3/8" x 1/8" Brass)	
S-4773	2	Bushing (1/2" x 1/4" Brass)	
S-4821	2	Ell (1/4" x 45° Brass)	
S-4822	1	Ell (1/2" x 90° Brass)	
S-5146	As	Copper Tubing (3/8" x 19 Gauge)	
S-5149	As	Copper Tubing (1/2" x 18 Gauge)	
S-5865	7	Connector (1/4" x 3/8")	
S-5868	6	Connector (3/8" x 1/2")	
S-5878	13	Elbow (1/4" x 3/8")	
S-5880	4	Elbow (3/8" x 1/2")	

SERVICE BULLETIN No. 313

THE AUTOCAR COMPANY, ARDMORE, PENNSYLVANIA
SERVICE DEPARTMENT

SUBJECT — WESTINGHOUSE BRAKE SERVICE CHART

CONDITION	CAUSE	REMEDY
Slow Pressure Build Up In Reservoirs	Leaking application or brake valve.	Clean valves or replace with reconditioned unit.
	Leaking compressor discharge valve.	Clean valve or replace head with reconditioned unit.
	Leaking lines or connections.	Replace tubing and fittings or tighten fittings.
	No clearance on unloader valves.	Adjust valve to .010" clearance.
	Clogged air cleaner.	Clean.
	Worn piston and rings, carbon in discharge line.	Replace with reconditioned unit.
Quick Loss of Reservoir Pressure When Motor Is Stopped	Worn and leaking compressor discharge valves.	Clean valves or replace head with reconditioned unit.
	Tubing or connections leaking.	Replace tubing or tighten fittings.
	Leaking valves.	Clean or replace unit.
	Leaking governor.	Clean or replace unit.
Compressor Not Unloading	Broken unloader diaphragm.	Install new diaphragm.
	Too much clearance on unloader valves.	Adjust to .010" clearance.
	Restriction in line from governor to unloader.	Replace tubing or clean.
	Governor not operating.	Replace with reconditioned unit.
Slow Brake Application	Low brake line pressure (Brake valve to chambers).	Adjust pressure through valve.
	Brake chamber push rod travel excessive.	Adjust brakes.
	Restriction in line.	Clean or replace tubing or hose.
	Leaking brake chamber diaphragm.	Replace diaphragm.
	Brake lining or Drum condition.	Replace or recondition.
	Leaking brake valve diaphragm.	Replace diaphragm or complete unit.
Slow Brake Release	Brake valve lever not returning fully to stop.	Adjust operating rod.
	Binding cam or cam shafts.	Lubricate and align properly.
	Brake chamber push rod travel excessive.	Adjust brakes.
	Restriction in tubing or hose.	Clean or replace.
	Improper seating of valves.	Clean or replace with a reconditioned unit.
Inefficient Brakes	Low brake line pressure.	Adjust pressure through brake valve.
	Excessive push rod travel on brake chambers.	Adjust brakes.
	Lining and drum condition.	Replace or repair.
	Brake chamber diaphragm leaking.	Replace diaphragm.

For more detailed information consult Instruction Pamphlet No. 5057

Genuine Westinghouse Air Brakes Maintenance Instructions

BENDIX-WESTINGHOUSE AUTOMOTIVE AIR BRAKE CO.

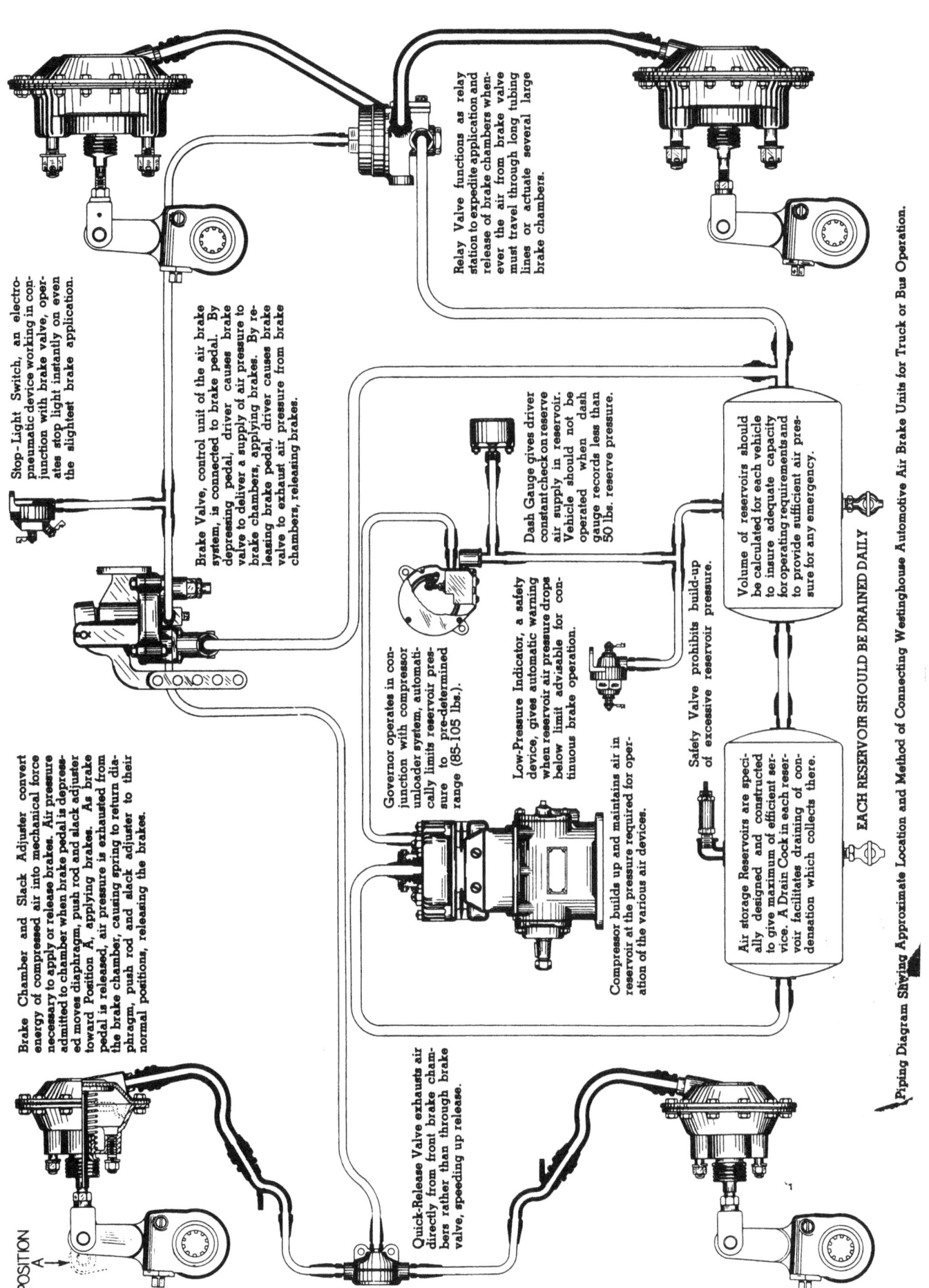

THE WESTINGHOUSE AUTOMOTIVE AIR BRAKE SYSTEMS

● Because of the unique flexibility of Air Control there is a Westinghouse Automotive Air Brake System available for every type of commercial operation. The basic devices involved in these systems are shown in the accompanying diagram and are also illustrated individually, together with a pictorial summary of inspection and maintenance necessary to obtain the maximum service from each device.

For synchronized tractor-trailer Air Brake operation trailer brake chambers and air lines of flexible hose are added and, generally, a relay-emergency valve is used to give an automatic trailer brake application in case of accidental breakaway. Independent Trailer Control helps to keep units stretched when operating under adverse traction conditions. The Automatic Emergency Equipment gives all units the protection of an automatic brake application if any line is broken or develops excessive leaks.

COMPRESSOR

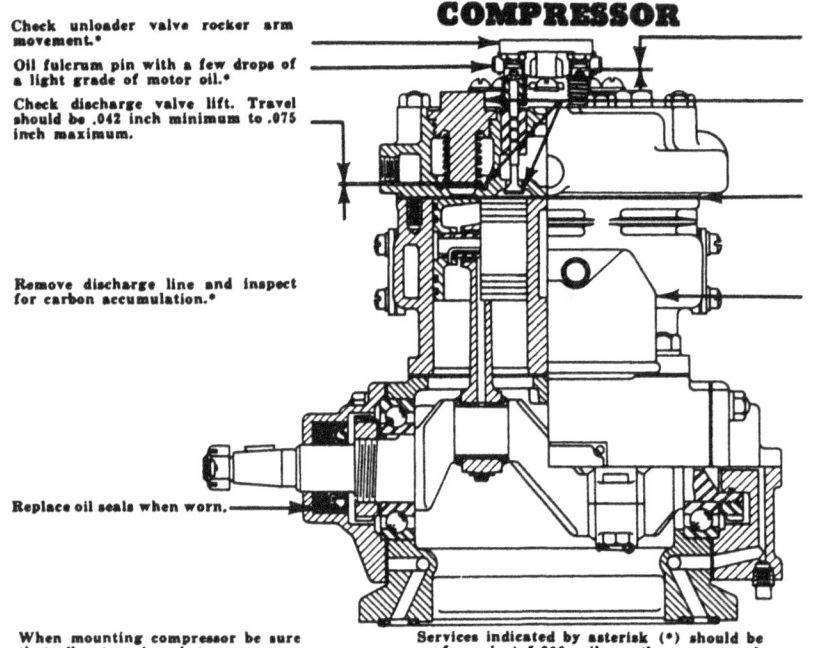

Check unloader valve rocker arm movement.*

Oil fulcrum pin with a few drops of a light grade of motor oil.*

Check discharge valve lift. Travel should be .042 inch minimum to .075 inch maximum.

Remove discharge line and inspect for carbon accumulation.*

Replace oil seals when worn.

When mounting compressor be sure that all external gaskets are new.

Check unloader valve clearance. This should be .010 inch minimum to .015 inch maximum.*

Remove cap nuts and check discharge valve seats for carbon. If carbonization is excessive, remove the cylinder head and clean carbon away from discharge and unloader valves, chambers and springs.*

When replacing cylinder head, be sure to use a new cylinder head gasket.*

At each 2,000 miles remove and dismantle strainer. Wash pulled curled hair with kerosene or gasoline. Blow it dry with a shop air line and saturate with a light lubricating oil before replacing.

ENGINE-LUBRICATED COMPRESSOR
Check oil pressures by installing gauge in oil line. Pressure should not be less than 5 pounds per square inch at idling speed nor less than 15 pounds per square inch at engine-governed speeds.*

SELF-LUBRICATED COMPRESSOR
Remove bayonet plug and check oil supply daily; replenish if necessary. Drain oil at same intervals motor crankcase oil is changed; flush out crankcase with clean oil and refill with best grade of motor oil. Complete oil specifications furnished upon request.

Services indicated by asterisk (*) should be performed at 5,000 miles; others as stated.

● The Compressor used with this system is a sturdily constructed 2- or 3-cylinder reciprocating type compressor — that is, its two or three pistons are moved up and down in separate cylinder bores by an engine-driven rotary crankshaft. Air compression is obtained when the pistons force air out of the cylinder bores, past the discharge valves and into the reservoir, gradually increasing the pressure of the air restrained in the reservoirs until it reaches the pressure desired for operation of the air brake devices.

The size of a compressor is rated in accordance with the number of cubic feet of free air displaced per minute while operating at a speed of 1,250 revolutions. Four sizes are used with the Westinghouse Automotive Air Brake System, namely 4, 6, 7¼ and 12 cubic feet.

Lubrication of the moving parts of the compressor is accomplished either by the compressor obtaining its oil supply directly from the oil system of the vehicle's engine or by a self-contained oil pumping system in the compressor.

GOVERNOR

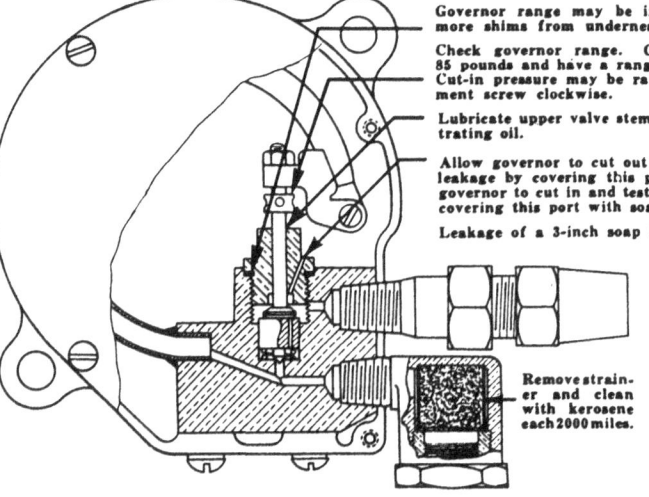

● The O-1 Governor is connected into the supply reservoir. As the reservoir pressure builds up to 105 pounds per square inch it deflects the tube of the governor, releasing tension from the upper valve. As this is done the air pressure from the reservoir is sufficient to open the lower valve and close the upper valve, blocking off the exhaust and permitting the air pressure to pass through the unloader line to the unloader chamber of the compressor. Then, as reservoir pressure is reduced to 85 pounds per square inch the tube again places sufficient tension on the upper valve to close the lower valve and open the upper valve, releasing unloader line pressure to atmosphere.

Governor range may be increased by taking one or more shims from underneath the upper valve guide.

Check governor range. Governor should cut in at 85 pounds and have a range of 15 to 20 pounds. The Cut-in pressure may be raised by turning the adjustment screw clockwise.

Lubricate upper valve stem with a few drops of penetrating oil.

Allow governor to cut out and check upper valve for leakage by covering this port with soap suds. Allow governor to cut in and test lower valve for leakage by covering this port with soap suds.

Leakage of a 3-inch soap bubble in 3 seconds is permissible. Leakage is caused either by dirt on the valve or on the valve seat, or by a badly worn valve. The leakage, if caused by dirt, can be remedied by cleaning both the valve and valve seat and then regrinding the valve slightly with Grade 1000 B-W Grinding Compound. If the leakage is caused by a worn valve the governor should be replaced with a reconditioned unit.

Remove strainer and clean with kerosene each 2000 miles.

WHERE NO MILEAGE BASIS IS GIVEN THE INSPECTION OF EACH DEVICE IS TO BE MADE A

Page 124

BRAKE CHAMBER and SLACK ADJUSTER

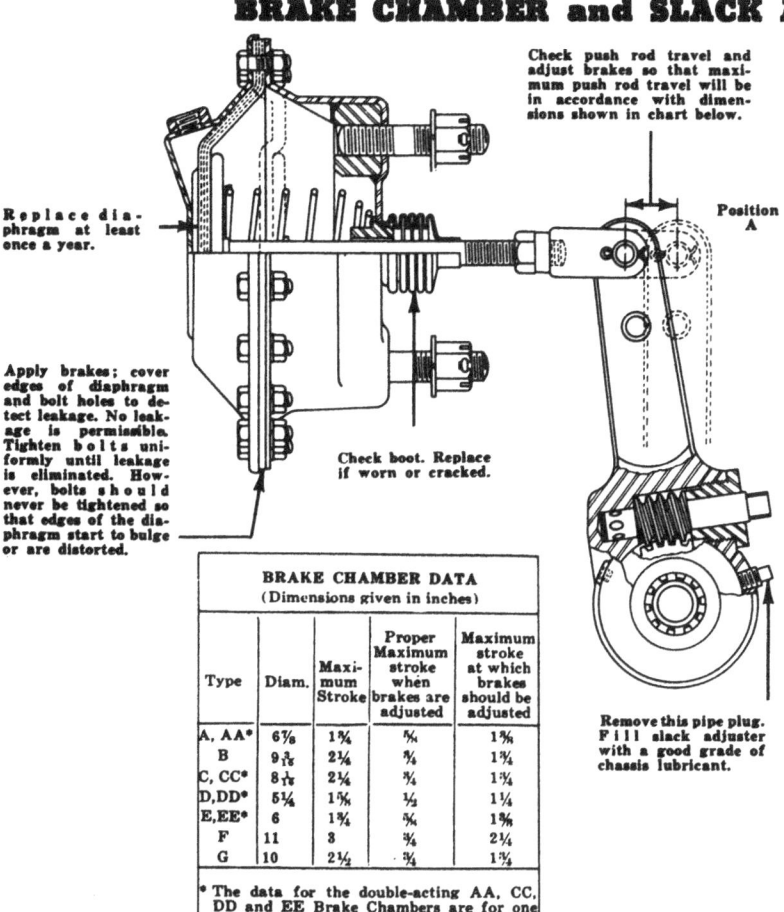

Replace diaphragm at least once a year.

Apply brakes; cover edges of diaphragm and bolt holes to detect leakage. No leakage is permissible. Tighten bolts uniformly until leakage is eliminated. However, bolts should never be tightened so that edges of the diaphragm start to bulge or are distorted.

Check push rod travel and adjust brakes so that maximum push rod travel will be in accordance with dimensions shown in chart below.

Check boot. Replace if worn or cracked.

Remove this pipe plug. Fill slack adjuster with a good grade of chassis lubricant.

Brake adjustments are made by turning this nut.

BRAKE CHAMBER DATA
(Dimensions given in inches)

Type	Diam.	Maximum Stroke	Proper Maximum stroke when brakes are adjusted	Maximum stroke at which brakes should be adjusted
A, AA*	6⅞	1¾	⅝	1⅜
B	9 3/16	2¼	¾	1¾
C, CC*	8 1/16	2 1/16	¾	1¾
D, DD*	5¼	1⅝	½	1¼
E, EE*	6	1¾	⅝	1 5/16
F	11	3	¾	2¼
G	10	2½	¾	1¾

* The data for the double-acting AA, CC, DD and EE Brake Chambers are for one side only.

● Air pressure admitted into the Brake Chamber when the brake pedal is depressed moves the diaphragm, push rod and slack adjuster toward position A, applying the brakes. As the brake pedal is released the air pressure is exhausted from the chamber and the spring returns the diaphragm, push rod and slack adjuster to their normal positions, releasing the brakes.

In normal braking the entire Slack Adjuster operates as a unit, rotating bodily with the cam shaft as the brakes are applied or released.

The most efficient brake action will be obtained when the slack adjuster arm travel is held to a minimum so that full length of the lever is used. The brake adjustments necessary to maintain proper slack adjuster arm travel are made by turning the adjusting worm. This rotates the worm gear, cam shaft and cam, expanding the brake shoes so that the slack caused by brake lining wear is taken up and the slack adjuster arm travel is returned to the minimum setting. **These brake adjustments usually average less than 5 minutes to a wheel, with Westinghouse slack adjusters.**

BRAKE VALVES

● The Brake Valve actually consists of two valves, the intake and the exhaust valves, which are alternately opened and closed. As the brake lever is moved toward the application position the graduating spring deflects the diaphragm downward, closing the exhaust valve and opening the intake valve, permitting the air pressure to pass from the reservoir to the brake chambers. When air pressure in the brake chambers and the cavity below the diaphragm is strong enough to balance the pressure placed on the graduating spring, it lifts the diaphragm and compresses the graduating spring so that both the intake and exhaust valves are held closed. Further lever movement toward the application position compresses the spring still further and causes the balance to be established at a higher pressure. Movement toward the release position removes pressure from the graduating spring, while air pressure lifts the diaphragm, permitting the exhaust valve to open and exhaust air from the brake chambers until a lower balance is reached. This balancing feature gives the driver a fine degree of brake control, enabling him to graduate the brake on or off as conditions demand.

All D Brake Valves are preloaded; the B 4 A Brake Valves are furnished either with or without the preloaded feature.

In preloaded valves the graduating spring is held compressed so that the first pedal movement causes the brake valve to deliver a predetermined pressure before graduation of air pressure begins. The preloading device

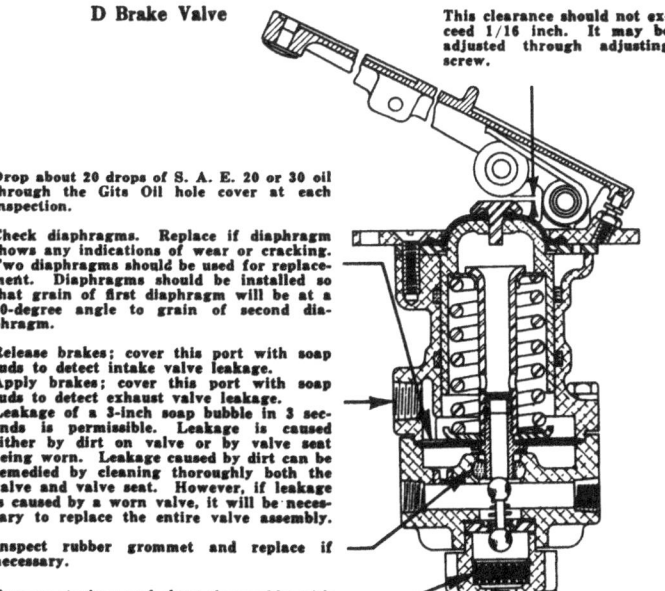

D Brake Valve

This clearance should not exceed 1/16 inch. It may be adjusted through adjusting screw.

Drop about 20 drops of S. A. E. 20 or 30 oil through the Gits Oil hole cover at each inspection.

Check diaphragms. Replace if diaphragm shows any indications of wear or cracking. Two diaphragms should be used for replacement. Diaphragms should be installed so that grain of first diaphragm will be at a 90-degree angle to grain of second diaphragm.

Release brakes; cover this port with soap suds to detect intake valve leakage.
Apply brakes; cover this port with soap suds to detect exhaust valve leakage.
Leakage of a 3-inch soap bubble in 3 seconds is permissible. Leakage is caused either by dirt on valve or by valve seat being worn. Leakage caused by dirt can be remedied by cleaning thoroughly both the valve and valve seat. However, if leakage is caused by a worn valve, it will be necessary to replace the entire valve assembly.

Inspect rubber grommet and replace if necessary.

Remove strainer and clean thoroughly with gasoline at least once every six months.

THE REGULAR PERIOD RECOMMENDED BY MANUFACTURER FOR THE VEHICLE'S INSPECTION.

Check with a test gauge for maximum pressure delivered to the brake chambers. If pressure varies 5 pounds from the original setting of the brake valve, the adjustment may be made by:
(1) Lengthening or shortening pedal rod.
(2) Moving clevis to a higher or lower hole in the brake valve lever.
Average maximum delivery pressure, 70 lbs.

Be sure that the lever stop is against the cover when the lever is in release position.

Oil lever pin with a few drops of lubricating oil.

Be sure that no strain is placed on the valve lever due to edge of cap hitting against cover when valve is in applied position.

Check diaphragms. Replace diaphragm if it shows any indication of wear or cracking. Two diaphragms should be used for replacement, installed so that the grain of first diaphragm is at a 90-degree angle to grain of second diaphragm.

Leakage from either intake or exhaust valve should not exceed a 3-inch bubble in 3 seconds. Leakage is caused by dirt, either on the valve or valve seat, or by a worn valve. Leakage caused by dirt can be remedied by cleaning both the valve and the valve seat, then regrinding valve with Bendix-Westinghouse Grade 1000 Grinding Compound. If leakage is caused by a worn valve the valve assembly should be replaced with a reconditioned unit.

Release brakes; cover exhaust port with soap suds to detect intake valve leakage. Apply brakes; cover exhaust port with soap suds to detect exhaust valve leakage.

ENLARGED VIEW OF TAPER-SEAT VALVES

Correct type of valve seat: Narrow seat will give better seal.

Incorrect type of valve seat: Wide or rounded seat will not seal. If valve seat is worn wider than 1/16 inch, a correct and efficient seat cannot be obtained.

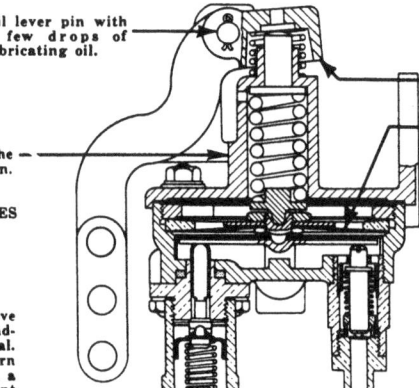

B 4 A Brake Valve

is usually set at just sufficient pressure to remove slack from the rigging, permitting efficient full pedal travel.

The D Brake Valve is furnished in three types, vertical, horizontal, and lever. The first two are mounted on the floor of the cab, with the treadle serving as the brake pedal. The third, mounted on the vehicle's frame, is connected to the standard brake pedal. In the D valve a simple ball type intake and exhaust valve assembly controls the flow of air pressure. A passage in the exhaust valve seat leads up through the diaphragm to the exhaust port. Downward treadle movement causes the exhaust valve seat to contact the upper ball of the intake and exhaust valve assembly, closing the exhaust passage, and moves the lower ball away from the intake valve seat, opening the intake valve. As this force is removed, a spring closes the intake valve.

In the B 4 A Brake Valves the application and release is obtained through the use of separate intake and exhaust valves, located entirely below the diaphragm. The rocker arm transfers force from the diaphragm to the valves, closing the exhaust valve and opening the intake valve. As the force is removed, springs open the exhaust valve and close the intake valve.

RELAY and RELAY-EMERGENCY VALVES

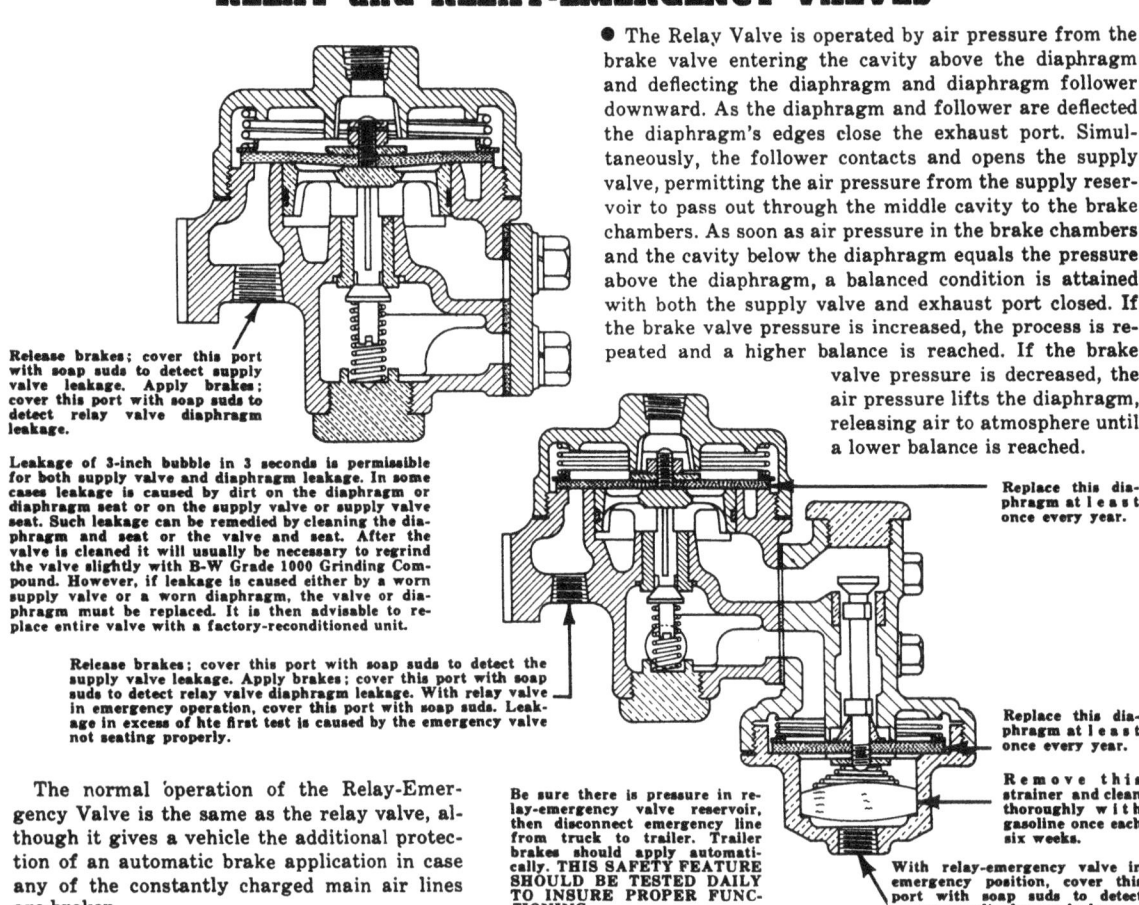

● The Relay Valve is operated by air pressure from the brake valve entering the cavity above the diaphragm and deflecting the diaphragm and diaphragm follower downward. As the diaphragm and follower are deflected the diaphragm's edges close the exhaust port. Simultaneously, the follower contacts and opens the supply valve, permitting the air pressure from the supply reservoir to pass out through the middle cavity to the brake chambers. As soon as air pressure in the brake chambers and the cavity below the diaphragm equals the pressure above the diaphragm, a balanced condition is attained with both the supply valve and exhaust port closed. If the brake valve pressure is increased, the process is repeated and a higher balance is reached. If the brake valve pressure is decreased, the air pressure lifts the diaphragm, releasing air to atmosphere until a lower balance is reached.

Release brakes; cover this port with soap suds to detect supply valve leakage. Apply brakes; cover this port with soap suds to detect relay valve diaphragm leakage.

Leakage of 3-inch bubble in 3 seconds is permissible for both supply valve and diaphragm leakage. In some cases leakage is caused by dirt on the diaphragm or diaphragm seat or on the supply valve or supply valve seat. Such leakage can be remedied by cleaning the diaphragm and seat or the valve and seat. After the valve is cleaned it will usually be necessary to regrind the valve slightly with B-W Grade 1000 Grinding Compound. However, if leakage is caused either by a worn supply valve or a worn diaphragm, the valve or diaphragm must be replaced. It is then advisable to replace entire valve with a factory-reconditioned unit.

Release brakes; cover this port with soap suds to detect the supply valve leakage. Apply brakes; cover this port with soap suds to detect relay valve diaphragm leakage. With relay valve in emergency operation, cover this port with soap suds. Leakage in excess of the first test is caused by the emergency valve not seating properly.

The normal operation of the Relay-Emergency Valve is the same as the relay valve, although it gives a vehicle the additional protection of an automatic brake application in case any of the constantly charged main air lines are broken.

Be sure there is pressure in relay-emergency valve reservoir, then disconnect emergency line from truck to trailer. Trailer brakes should apply automatically. THIS SAFETY FEATURE SHOULD BE TESTED DAILY TO INSURE PROPER FUNCTIONING.

Replace this diaphragm at least once every year.

Replace this diaphragm at least once every year.

Remove this strainer and clean thoroughly with gasoline once each six weeks.

With relay-emergency valve in emergency position, cover this port with soap suds to detect emergency diaphragm leakage.

WHERE NO MILEAGE BASIS IS GIVEN THE INSPECTION OF EACH DEVICE IS TO BE MADE

SAFETY VALVE

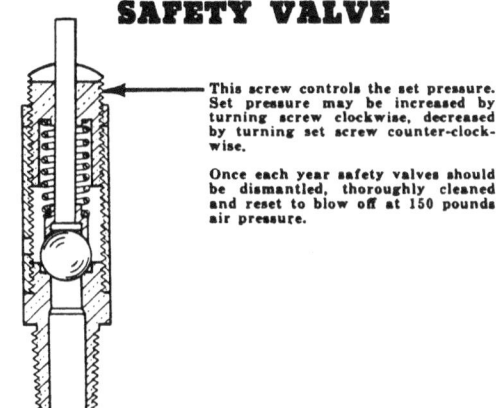

This screw controls the set pressure. Set pressure may be increased by turning screw clockwise, decreased by turning set screw counter-clockwise.

Once each year safety valves should be dismantled, thoroughly cleaned and reset to blow off at 150 pounds air pressure.

• The Safety Valve is connected into the pilot reservoir. When reservoir pressure is built up to exceed 150 pounds per square inch, the air pressure forces the ball off its seat, releasing pressure in excess of 150 pounds to atmosphere. When the pressure drops below 150 pounds the spring forces the ball back on its seat.

STOP-LIGHT SWITCH

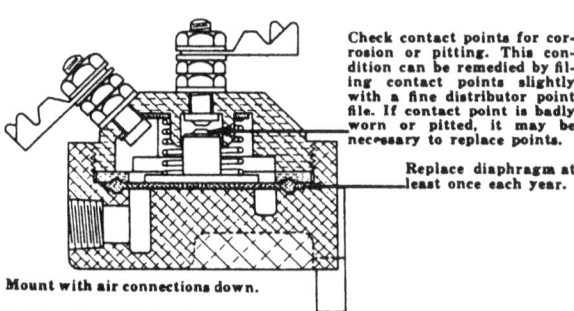

Check contact points for corrosion or pitting. This condition can be remedied by filing contact points slightly with a fine distributor point file. If contact point is badly worn or pitted, it may be necessary to replace points.

Replace diaphragm at least once each year.

Mount with air connections down.

• The Stop-Light Switch also is operated by air pressure in the cavity acting upon the diaphragm above. As pressure is admitted to this cavity the diaphragm is raised, establishing contact between the terminals and completing an electrical circuit. As the air pressure is exhausted from this cavity the spring depresses the diaphragm, breaking the contact.

QUICK-RELEASE VALVE

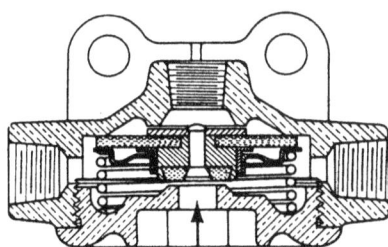

Apply brakes; cover this port with soap suds to detect leakage. Leakage is caused either by dirt on the diaphragm exhaust valve seat or by a worn exhaust valve seat. Dirt leakage may be remedied by cleaning the exhaust valve seat. However, if leakage is caused by a worn exhaust seat, replace entire diaphragm assembly.

• The Quick-Release Valve is operated by air pressure from the brake valve entering into the cavity above the diaphragm. As the air pressure enters this cavity it forces the exhaust seat of the diaphragm against the edges of the exhaust port, sealing the exhaust and permitting the air pressure from the brake valve to pass to the brake chambers. As the brake valve is released the air pressure and the force of the spring lift the diaphragm, removing the exhaust seat from the exhaust port and releasing the air pressure accumulated in the brake chambers to atmosphere.

DASH GAUGE

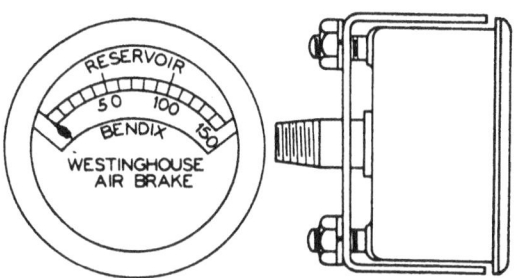

Check with an accurate Westinghouse test gauge. If dash gauge varies 4 pounds it should be replaced with a reconditioned unit.

• The Dash Gauge is connected directly to the supply reservoir and constantly registers reservoir pressure. The gauge mechanism is contained in a sealed case and there is no method for servicing it other than to return the gauge on the regular repair-exchange basis.

LOW-PRESSURE INDICATOR

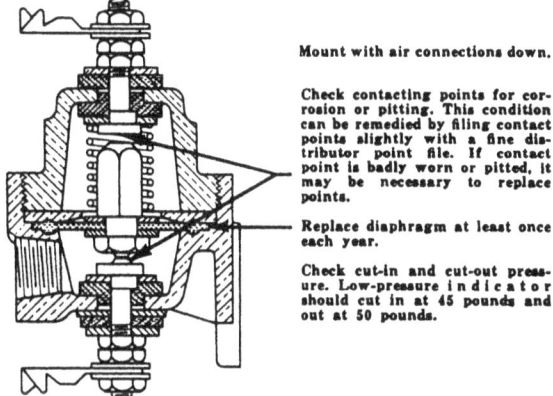

Mount with air connections down.

Check contacting points for corrosion or pitting. This condition can be remedied by filing contact points slightly with a fine distributor point file. If contact point is badly worn or pitted, it may be necessary to replace points.

Replace diaphragm at least once each year.

Check cut-in and cut-out pressure. Low-pressure indicator should cut in at 45 pounds and out at 50 pounds.

The Low-Pressure Indicator is operated by air pressure passed directly from the supply reservoir into the cavity below the diaphragm. When air pressure in this cavity is below 50 pounds per square inch the spring holds the diaphragm down, maintaining a contact between the terminals and establishing an electrical circuit. When the air pressure exceeds 50 pounds the diaphragm is raised, breaking the contact.

FACTORY REPAIR-EXCHANGE

• To enable the operator to obtain the maximum efficiency from his air brakes at the most economical maintenance costs, the Bendix-Westinghouse Automotive Air Brake Company has established the Repair-Exchange Service. By this plan the operator, upon payment of a standard flat rate charge, may exchange any service-worn air brake unit for a genuine factory-reconditioned unit at the nearest Bendix-Westinghouse distributor.

Each distributor maintains a large stock of reconditioned units, which enables the operator to replace any worn air brake device without interrupting the vehicle's service while the unit is being repaired.

All reconditioned units are covered by the identical guarantees of a new air brake unit and will give the same dependable service. Since these worn air brake devices are reconditioned by the standard original production methods, the Bendix-Westinghouse Automotive Air Brake Company is able to establish the flat rate price at the most economical point.

THE REGULAR PERIOD RECOMMENDED BY MANUFACTURER FOR THE VEHICLE'S INSPECTION.

SERVICE BULLETIN No. 400

THE AUTOCAR COMPANY, ARDMORE, PENNSYLVANIA
SERVICE DEPARTMENT

SUBJECT   BUDD WHEELS   -   APPLICATION AND MAINTENANCE

RIGHT AND LEFT HAND THREADS are used on all BUDD DUAL assemblies to insure cap nuts staying tight. In mounting hubs on chassis, or in replacing studs or nuts, right-hand studs must be used on right side of chassis and left-hand on left side. RIGHT and LEFT sides are seen by the driver facing forward. All studs and nuts are plainly marked "R" and "L" and must be so used.

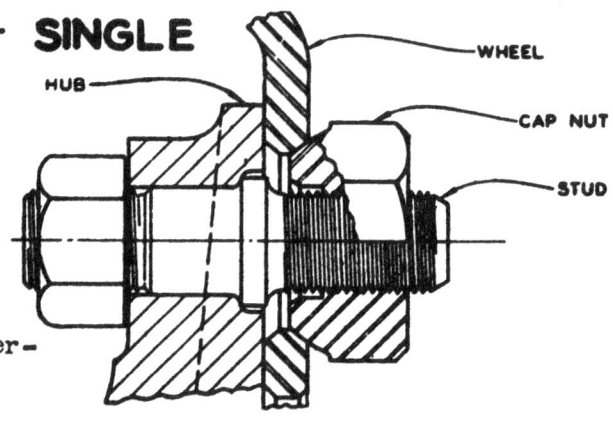

FIG. 1

WHEELS MUST BE CLEAN. Always examine wheels before mounting on hub to be sure that countersunk holes where ball face of cap nut seats, are free from dirt or paint and that face of disc and hub flange, where they bear together, is clean and free from dirt, grease or paint. The same applies to the surfaces of both rear duals where the two come together. Presence of foreign matter will prevent proper bearing and create high spots which are likely to cause loose fits, play and wear. Watch this point, particularly in mounting spare wheels, which may have picked up road dirt. The countersunk holes should be carefully cleaned when a wheel is painted.

ALL BUDD DUAL REAR WHEELS now produced are of the double cap nut type as illustrated in Fig. 2. The inner dual wheel is individually held by the sleeve-shaped inner cap nut, and in applying the wheel must be mounted and tightened before the outer wheel is put on. The outer wheel slips over the inner cap nuts, and is independently held by the outer nuts. The front, or single wheel, is held by a single set of nuts (Fig. 1)

TIGHTENING NUTS should be done with the truck jacked up. Re-tighten all cap nuts after running approximately 50 miles under load after the first installation, or after wheel change. OUTER CAP NUTS MUST BE BACKED OFF AT LEAST TWO FULL TURNS TO TIGHTEN INNER NUTS WHICH MUST NOT BE NEGLECTED. In mounting wheels or tightening nuts, proceed in a criss-cross fashion and not around the circle. DO NOT USE AN EXTENSION IN THE REGULAR WRENCH HANDLE AS SUPPLIED. Ordinary pressure as exerted in tightening cap nuts with the handle is sufficient to drive cap nuts home without use of an extension.

AN OCCASIONAL CHECK of nuts for tightness is desirable, especially soon after a tire change has been made on the road. Properly installed, they should remain tight indefinitely.

MOUNT WHEELS with valve stems opposite, whether disc or spoke type, to permit easy inflation of tires.

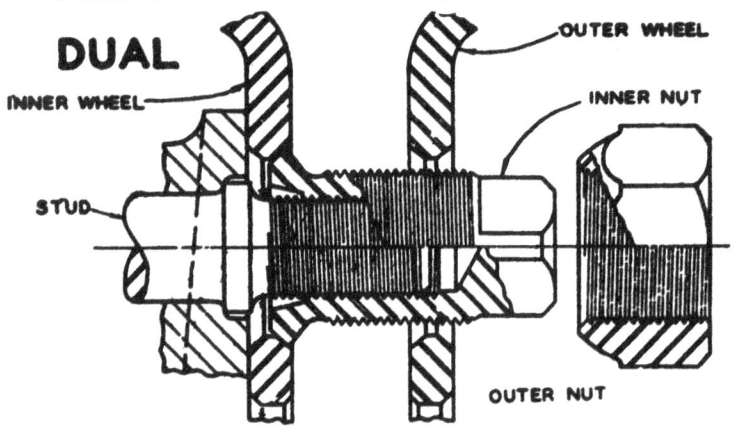

ABOVE SHOWS ORDER OF ASS'Y - TIGHTEN INNER WHEEL SECURELY TO HUB BEFORE APPLYING OUTER WHEEL

FIG. 2

PARTS LIST

No. 857

THE AUTOCAR COMPANY, ARDMORE, PENNSYLVANIA
SERVICE DEPARTMENT

BULLETIN REF.

ASSEMBLY WHEELS & BEARINGS - MODELS U-2044 & U-4044 - U.S.A. 1940

Part No.	Req.	Description	Ref. No.
9BJ0403	2	Front Wheel	Budd #33287
9BP004	2	Front Wheel Hub	
4FD0138	2	Front Wheel Bearing Cone-Inner	Timken #594
4FD0139	2	Front Wheel Bearing Cup-Inner	Timken #592A
19SKB0445	2	Front Wheel Bearing Cone-Outer	Timken #498
19SKB0444	2	Front Wheel Bearing Cup-Outer	Timken #493
9ZDK028	2	Front Wheel Bearing Adjusting Nut	Timken #1827-H-34
9ZDK0175	2	Front Wheel Bearing Lock Washer	Timken #1829-G-85
9ZDK028	2	Front Wheel Bearing Lock Nut	Timken #1827-H-34
9ZDK095	2	Front Wheel Bearing Felt Washer	Timken #5-X-285
9BP0809	2	Front Hub and Drum	
9BP0232	2	Front Brake Drum	
4A3489	10	Front Wheel Studs - R.H.	Budd #12247
4A3491	10	Front Wheel Studs - L.H.	Budd #12248
4A0499	20	Front Wheel Stud Nut	Budd #13332
9BJ0403	4	Rear Wheel	Budd #33287
4BP004	2	Rear Wheel Hub	
4UG0559	2	Rear Wheel Spacer	
4UG0141	2	Rear Wheel Bearing Cone-Inner	Timken #566
4Y2211	2	Rear Wheel Bearing Cup-Inner	Timken #563
4Y2141	2	Rear Wheel Bearing Cone-Outer	Timken #560
4NFG0142	2	Rear Wheel Bearing Cup-Outer	Timken #552A
4D0550	2	Rear Wheel Bearing Adjusting Nut	Timken #AT-6880
4D0166	2	Rear Wheel Bearing Lock Washer	Timken #T-3840
4D025	2	Rear Wheel Bearing Lock Nut	Timken #T-3564
9DKA095	2	Rear Wheel Bearing Felt Washer	Timken #5-X-282
4BP0809	2	Rear Hub and Drum	
4BP021	2	Rear Brake Drum	
4UK3489	10	Rear Wheel Studs - R.H.	Budd #18309
4UK3491	10	Rear Wheel Studs - L.H.	Budd #18310
4A0499	20	Rear Wheel Stud Nut	Budd #13332

www.ingramcontent.com/pod-product-compliance
Lightning Source LLC
LaVergne TN
LVHW061311060426
835507LV00019B/2106